The Connectivity Hypothesis

The Connectivity Hypothesis

Foundations of an Integral Science of Quantum, Cosmos, Life, and Consciousness

ERVIN LASZLO

Foreword by
Ralph H. Abraham

STATE UNIVERSITY OF NEW YORK PRESS

Published by
State University of New York Press, Albany

© 2003 State University of New York

All rights reserved

Printed in the United States of America

No part of this book may be used or reproduced in any manner whatsoever without written permission. No part of this book may be stored in a retrieval system or transmitted in any form or by any means including electronic, electrostatic, magnetic tape, mechanical, photocopying, recording, or otherwise without the prior permission in writing of the publisher.

For information, address State University of New York Press,
194 Washington Avenue, Suite 305, Albany, NY 12210-2384

Production by Diane Ganeles
Marketing by Michael Campochiaro

Library of Congress Cataloging-in-Publication Data

Laszlo, Ervin, 1932–
    The connectivity hypothesis : foundations of an integral science of quantum, cosmos, life, and consciousness / Ervin Laszlo ; foreword by Ralph H. Abraham.
      p. cm.
    Includes bibliographical references and index.
    ISBN 0-7914-5785-0 (alk. paper) — ISBN 0-7914-5786-9 (alk. paper)
    1. Science—Philosophy. 2. Cosmology. I. Title.

Q175 .L2854 2003
501—dc21   2002030480

10 9 8 7 6 5 4 3 2 1

Contents

Foreword — vii

Introduction — 1

PART 1 Coherence in Nature and Mind — 3

1. Coherence in the Physical World — 5
2. Coherence in the Living World — 17
3. Coherence in the Sphere of Mind — 27
4. Understanding Coherence: The Elements of an Explanation — 39

PART 2 The Connectivity Hypothesis — 49

5. Premises — 51
6. Postulates — 65
7. The Hypothesis — 73
8. Coherence Explained: Testing the Power of the Hypothesis — 79
9. The Advent of Integral Quantum Science — 95

Postscript — 103

    The Metaphysics of Connectivity — 103

    Philosophical Implications — 110

Appendix 1. General Relativity and the Physical Vacuum 119
Reconsidering Einstein's Equations in Relation to Connectivity Hypothesis
László Gazdag

Appendix 2. Healing through the Ψ Field:
Two Experiments 125
Maria Sági

References 133

Index 143

Foreword

Pythagoras, ancient theologian and prophet, imagined a model for cosmos and consciousness based on number mysticism. But two thousand years would pass before his prophecy could be realized. Then in 1637, René Descartes and Pierre de Fermat independently connected algebra and geometry, an adequate basis for modern science. And very swiftly we had Galileo, Isaac Newton, Jean le Rond d'Alembert, Jean Baptiste Fourier, James Clerk Maxwell, Albert Einstein, Erwin Schrödinger, and the field theories of mathematical physics—scalar, vector, tensor, spinor fields, and so on. The attempts to unify all the fields into a single mathematical model began with Einstein and are ongoing today. The current state of the art, known as the theory of the quantum vacuum field, attempts to model the wholeness and connectedness of the physical universe, from quantum to cosmos. Meanwhile, the methods and dreams of mathematical physics were applied to biology by Nicholas Rashevsky, and to psychology by Kurt Lewin, in the 1930s.

More recently, hopes grow for a science of consciousness, and many capable scientists are engaged in experimental and theoretical work aimed at models inspired by the field theories of mathematical physics, especially quantum theory.

Somehow, and nobody knows quite why, the upper, vital spheres of the perennial philosophy—intellect, soul, and spirit—were dropped out of the picture, as modern science took over from philosophy and religion at the end of the Renaissance. The new science of life of Rupert Sheldrake tries to restore vitalism to biology. The archetypal psychology of Carl Jung, James Hillman, and Thomas Moore tries to bring it back into psychology. Along with many others, these efforts may be seen as a New Renaissance.

Amid the milieu of this embryonic paradigm shift, Ervin Laszlo stands out as the unique champion of a holistic philosophy of the broadest perspective. For his bold plan is to unify all—quantum, cosmos, life, and consciousness—in a single grand unified model. In this book he summarizes the new empirical results that now trigger a paradigm shift; details his blueprint for the conceptual foundations of a unified theory of quantum, cosmos, life, and consciousness; and works out the implications of the new theory for the outstanding philosophical problems unresolved by the current paradigm. A major characteristic feature of the metaphysics inspired by his connectivity hypothesis is its bipolar aspect: the manifest domain of matter and the quantum vacuum, a cosmic plenum of infinite energy are in an endless loop of coevolution.

When a great grand unified theory will appear it will very likely conform to the prophetic vision of Ervin Laszlo. In this book, he points the way to an integral science of cosmos and consciousness, and provides the conceptual foundations for it: the hypothesis of connectivity. He puts before us the essential elements of the emerging paradigm of science in the twenty-first century.

Ralph H. Abraham
University of California at Santa Cruz

Introduction

In the first decade of the twenty-first century the main branches of the empirical sciences face a paradigm shift as deep as that which occurred at the beginning of the twentieth century, when classical physics gave way to relativity and subsequently to quantum physics. The current shift goes beyond the ruling paradigm of twentieth century science, to a new and different paradigm, more adequate to the facts that are now coming to light.

The paradigm-shift is triggered by a number of surprising observational and experimental findings; these do not fit into the established theories, or do so only at the cost of introducing arbitrary assumptions and auxiliary hypotheses. Maintaining the dominant paradigm in the face of the new evidence threatens the coherence of the scientific world picture—the very opposite of the phenomena that scientists are now called upon to explain. For the pertinent findings speak of a hitherto unsuspected form and level of coherence in nature. This kind of coherence means a quasi-instantaneously synchronized state, with nonconventional connections between the parts that make up a system, and between the systems and their environment. Such connections seem to obtain over all finite distances and finite times, and they suggest that the "nonlocality" discovered in the microscopic domain of the quantum may extend into the macroscopic domains of life, mind, and cosmos. Nature, it appears, is made up as a nested hierarchy of nonlocally connected coherent systems.

Time- and space-invariant coherence in diverse domains of observation and experiment calls for fundamentally new assumptions about the nature of reality. This book presents a hypothesis that responds to the findings. It suggests that space is not a vacuum but a plenum, and information, as physically effective "in-formation," is as fundamental as energy, and is likewise conserved. These concepts serve as a foundation for "integral quantum science," a transdisciplinary unified theory that furnishes the essential element of the paradigm that will ground science in the twenty-first century.

PART 1

Coherence in Nature and Mind

The finding of coherence at various scales of size and complexity in nature, from quanta, the smallest identifiable units of the physical world, to galactic macrostructures, the largest units, is not the finding of the standard form of coherence. The standard form is observed in optical interference experiments. Ordinary light sources are coherent—show an interference pattern—over a distance of a few meters, since the phase coherence of radiation from the same source lasts only 10 nanoseconds. Lasers, microwaves, and other technological light sources remain coherent for considerably longer and hence over greater distances. But the kind of coherence now coming to light is more complex and remarkable than the standard form, even if in this form, too, phase relationships remain constant and processes and rhythms are harmonized. The pertinent kind of coherence shows a quasi-instant correlation of the parts or elements of a system in space as well as time. All parts of a system of this coherence are correlated in such a way that what happens in and to one of the system's parts also happens in and to all its other parts, and hence it happens in and to the system as a whole. In consequence the parts respond to the "rest of the world" as a whole, maintain themselves as a whole, and change and evolve as a whole. This kind of coherence also obtains in the sphere of mind. It is recognized in quantum physics but has no realistic explanation, and it is mainly anomalous at the macroscopic level: the current paradigm of local action and interaction cannot account for it.

We begin our search for the integral science that would unify our understanding of quantum, cosmos, life, and consciousness by examining this remarkable form of coherence in nature and reviewing the pertinent evidence. In the four chapters that follow we describe the coherence found in quantum physics, physical cosmology, the biological world, and the emerging field of consciousness research, before discussing, in chapter 5, the kind of concepts that could furnish an explanation of it.

Chapter 1

✦

Coherence in the Physical World

Microscale Coherence: The Phenomenon of Quantum Nonlocality

In the physical world the anomalous form of coherence has been much researched and discussed. It is quantum coherence—the coherence among the quantized packets of matter-energy known as quanta.

The curious behavior of quanta is legendary. The light and energy quanta that come to light in the famous physics experiments do not behave as the small-scale equivalents of familiar objects. Until an instrument or an act of observation registers them, they have neither a unique location nor a unique state.

The state of the quantum is defined by the wavefunction that encodes the superposition of all the potential states the quantum can occupy. When the quantum is measured or observed, this superposed wavefunction collapses into the determinate state of a classical particle. Until then the quantum has the properties of both waves and corpuscles, that is, wave-particle complementarity. And, as Werner Heisenberg's indeterminacy principle indicates, its various properties cannot be measured at the same time. When one property is measured, the complementary property becomes blurred, or its value goes to infinity.

The superposed state of the quantum defies realistic explanation. This state obtains between one deterministic quantum state and another in the absence of observation, measurement, or another interaction. This period in time—which varies from a millisecond in the case of a pion decaying into two photons, a uranium atom decaying after ten thousand years, to a photon that may reach the retina of the eye of a human observer after eleven billion years—is regarded as one tick of a fundamental quantum clock, or q-tick. According to the standard Copenhagen interpretation, reality does not exist during a q-tick, only at the end of it, when the wavefunction has collapsed and the quantum has transited from the superposed indeterminate to a classical determinate state.

It is not clear, however, what process brings about the collapse of the wavefunction. Eugene Wigner speculated that it is due to the act of observation: the consciousness of the observer interacts with the particle. Yet also the instrument through which the observation is made could impart the crucial impetus, in which case the transition occurs whether or not an observer is present. Heisenberg affirmed now the former view, now the latter (Heisenberg 1955, 1975).

That the wavefunction of particles would collapse upon interaction has been demonstrated in experiments first conducted by Thomas Young in the early nineteenth century. Young made coherent light pass through an intervening surface with two slits. He placed a screen behind the intervening surface in order to receive the light penetrating the slits. Then a wave-interference pattern appears on the screen. One explanation is that photons have the property of waves: they pass through both slits. This becomes problematic when the light source is so weak that only one photon is emitted at a time. Such a single packet of light energy should be able to pass only through one of the slits. Yet, when seemingly corpuscular photons are emitted one after another, an interference pattern builds up on the screen, and this could only occur if the photons are waves.

In a related experiment by John A. Wheeler, photons are likewise emitted one at a time; they are made to travel from the emitting gun to a detector that clicks when a photon strikes it (Wheeler 1984). A half-silvered mirror is inserted along the photon's path; this splits the beam, giving rise to the probability that one in every two photons passes through the mirror and one in every two is deflected by it. To confirm this probability, photon counters that click when hit by a photon are placed both behind this mirror, and at right angles to it. The expectation is that on the average one in two photons will travel by one route and the other by the second route. This is confirmed by the results: the two counters register a roughly equal number of clicks—and hence of photons. When a second mirror is inserted in the path of the photons that were undeflected by the first, one would still expect to hear an equal number of clicks at the two counters: the individually emitted photons would merely have exchanged destinations. But this expectation is not borne out by the experiment. Only one of the two counters clicks, never the other. All the photons arrive at one and the same destination.

It appears that the photons interfere with one another as waves. Above one of the mirrors the interference is destructive—the phase difference between the photons is one hundred eighty degrees—so that the photon waves cancel each other. Below the other mirror the interference is constructive: the wave-phase of the photons is the same and as a consequence they reinforce one another.

Photons that interfere with each other when emitted moments ago in the laboratory also interfere with each other when emitted in nature at considerable intervals of time. The "cosmological" version of Wheeler's experiment bears witness to this. In this experiment the photons are emitted not by an artificial light source, but by a distant star. In one experiment the photons of the light beam emitted by the double quasar known as $0957 + 516A,B$ were tested. This distant quasi-stellar object is believed to be one star rather than two, the double image due to the deflection of its light by an intervening galaxy situated about one fourth of the distance from Earth. (The presence of mass, according to relativity theory, curves space-time and hence also the path of the light beams that propagate in it.) The deflection due to this "gravitational lens" action is large enough to bring together two light rays emitted billions of years ago. Because of the additional distance traveled by the photons that are deflected by the intervening galaxy, they have been on the way fifty thousand years longer than those that came by the more direct route. But, although originating billions of years ago and arriving with an interval of fifty thousand years, the photons interfere with each other just as if they had been emitted seconds apart in the laboratory.

It turns out that, whether photons are emitted at intervals of a few seconds in the laboratory, or at intervals of thousands of years in the universe, those that originate from the same source interfere with each other.

The interference of photons and other quanta is extremely fragile: any coupling with another system destroys it. Recent experiments indicate that when any part of the experimental apparatus is coupled with the source of the photons, the fringes that record the interference vanish. The photons behave as classical particles.

For example, in experiments designed to determine through which of the slits a given photon passes, a "which-path detector" is coupled to the emitting source. As a result the fringes weaken and ultimately vanish, indicating interference. The process can be calibrated: the higher the power of the which-path detector, the more of the fringes disappears. The experiment conducted by Mordechai Heiblum, Eyal Buks, and their colleagues at the Weizmann Institute in Israel made use of a device less than one micrometer in size, which creates a stream of electrons across a barrier on one of two paths (Buks et al. 1998). The paths focus the electron streams and enable the investigators to measure the level of interference between the streams. With the detector turned on for both paths, the interference fringes disappear as expected. But the higher the detector is tuned for sensitivity, the less interference patterns there are.

It appears that a physical factor enters into play: the coupling of the measuring apparatus to the light source. This coupling is closer than one would expect: in some experiments the interference fringes disappear as soon as the detector apparatus is readied—even when the apparatus is not turned on. Leonard Mandel's optical-interference experiment bears this out (Mandel 1991). In Mandel's experiment two beams of laser light were generated and allowed to interfere. When a detector is present that enables the path of the light to be determined, the fringes disappear. They disappear regardless of whether or not the determination is actually carried out. It appears that the very possibility of "which-path-detection" destroys the superposed-state of the photons.

This finding was confirmed in experiments carried out in 1998 at the University of Konstanz (Dürr et al. 1998). In these experiments the puzzling interference fringes were produced by the diffraction of a beam of cold atoms by standing waves of light. When there is no attempt to detect which path the atoms are taking, the interferometer displays fringes of high contrast. However, when information is encoded within the atoms as to the path they take, the fringes vanish. Yet the instrument itself cannot be the cause of the collapse—it does not deliver a sufficient "momentum kick." The back action path of the detector is four orders of magnitude smaller than the separation of the interference fringes. In any case, for the inference pattern to disappear the labeling of the paths does not need to be actually determined by the instrument: it is enough that the atoms are labeled so that the path they take *can* be determined.

These experiments can be performed whether or not anyone is watching; consequently they do away with the theory that a conscious observer is needed to collapse the wavefunction. And they also show that measurable physical interaction is not a necessary condition of the collapse: it also occurs in its absence.

A similar kind of intrinsic correlation among particles comes to light in the so-called EPR (Einstein-Podolski-Rosen) thought-experiment put forward in 1935 (Einstein, Podolski, Rosen 1935). In this experiment a particle is split in two, and the two halves are allowed to separate and travel a finite distance. Then a measurement is made of one aspect of the quantum state of one of the halves—such as the spin state—and a measurement of another aspect of the state of the other. Einstein proposed that since the quantum states of the particles are identical, we would then know both aspects of their state at the same time. This would show that the Heisenberg indeterminacy principle does not yield a complete description of physical reality.

When in the 1980s experimental apparatus sophisticated enough to test Einstein's thought experiment became available, it turned out that measuring, for example a spin component of particle A has an instantaneous effect on particle B: it causes B's spin wavefunction to collapse into a state with the opposite spin component (the permissible spin states are "up" or "down" along axes x, y, and z). Particle B manifests different states when different measurements are made on particle A—the effect depends on just what is measured on A. Thus the measurement on A does not merely reveal an already established state of B: it actually *produces* that state. Somehow, A "knows" when B is measured, and with what result, for it assumes its own state accordingly.

There appears to be a nonlocal connection between particles A and B. Empirical experiments first performed in the early 1980s by Alain Aspect and collaborators, and frequently repeated since then, show that this connection is intrinsic to the particles, and is not due to signals transmitted by the measuring apparatus (Aspect et al. 1982, Aspect & Grangier 1986, Selleri 1988, Duncan & Kleinpoppen 1988, Hagley et al. 1997, Tittel et al. 1998). The experiments involved more particles over ever-larger distances, without modifying these results. It appears that separation does not divide particles from each other. It is not necessary that the particles should have originated in the same quantum state; experiments show that any two particles, whether electrons, neutrons, or photons, can originate at different points in space and in time—they remain correlated as long as they had once assumed the same quantum state, that is, were part of the same coordinate system.

The results can be extrapolated to show that the correlations between quanta are invariant in regard to distance and time. Quanta that at one time and one place occupied the same quantum state can be light years apart in space and thousands of years apart in time, and still remain correlated.

Space- and time-transcending correlations are not explained by the assumption that a finite-velocity (even if supraluminal) signal would connect the particles. The quantum state appears to be intrinsically nonlocal. Already in his 1935 assessment of the EPR experiment Schrödinger maintained that particles in the quantum state do not have individually defined states: their states are fundamentally "entangled" with each other. The state of collective superposition applies to two or more properties of a single particle, the same as to a set of several particles. It is not the single particle or the single property of a particle that carries information on the quantum state, but the collective wavefunction of the system of coordinates in which the particles participate.

A mathematical specification of the collective state of particles within a given quantum system was furnished by Ke-Hsueh Li of the Chinese Academy of Sciences (Li 1992, 1994, 1995). He has shown that the Heisenberg uncertainty principle is an alternative approach to grasp the coherence properties of fields and particles. According to Li, interference between different probability amplitudes, and hence the coherence property of probability packets, must be understood in reference to "coherence space-time." Coherence time is the time within which interference between the packets exists, and coherence length (or volume) is the space within which such interference occurs. Coherence space corresponds to the breadth of the wave function which is the region within which matter (more exactly, matter-fields) and radiation (force-fields) are statistically distributed. Interference patterns are formed only within coherence space-time; beyond it, phase information is lost. Within coherence space-time supraluminal velocities can occur and nonlocality is the rule. Particles and fields constitute one indivisible whole.

Although the nature of nonlocality and entanglement are not yet definitively determined, it is already clear that these phenomena exist and make for a remarkable space- and time-transcending form of coherence among quanta. The quantum world is entirely hallmarked by this coherence—a major element in what Richard Feynman dubbed the "central mystery" of physics.

Macroscale Coherence: The Phenomenon of Cosmic Nonlocality

The kind of coherence observed in the domain of the quantum was believed to be limited to that domain; the world of macroscopic objects was thought to be "classical." Yet this assumption is no longer entirely true. There is growing evidence that an anomalous form of coherence also occurs at macroscopic scales; indeed, even at cosmic scales. The whole universe, it appears, has coherence-features that suggest that it is nonlocal (Nadeau 1999). The standard model of the universe, the cosmology of the Big Bang, cannot account for this finding.

Big Bang cosmology maintains that the universe originated in an explosive instability in the quantum vacuum. A region of this pre-space exploded, creating a fireball of staggering heat and density. In the first few milliseconds it synthesized all the matter that now populates space-time. The particle-antiparticle pairs that emerged from the vacuum collided with and annihilated each other; and the onebillionth of the originally created particles that survived (the tiny excess of particles over antiparticles) made up the matter-content of the universe we now observe. After about two hundred thousand years these particles decoupled from the radiation field of the primordial fireball: space became transparent, and clumps of matter established themselves as distinct elements of the cosmos. Due to gravitational attraction they condensed into gigantic swirls that solidified as galaxies. In time these became further structured as stars and stellar systems.

The overall features of Big Bang theory's "standard scenario" are well established; the computer analysis of some three hundred million observations made by NASA's Cosmic Background Explorer satellite (COBE) in 1991 provided confirmation. Detailed measurements of the cosmic microwave background radiation—the presumed remnant of the Big Bang—show that the variations derive from the original explosion and are not distortions caused by radiation from stellar bodies. They are the remnants of minute fluctuations within the cosmic fireball when it was less than one trillionth of a second old. They indicate the amount—if not the nature—of the particles of matter that were created (and not quasi-immediately annihilated) in the universe. If the surviving particles make for a matter-density above a certain number (estimated at 5×10^{-26} g/cm^3), the gravitational pull associated with the total amount of matter will ultimately exceed the inertial force generated by the Big Bang and the universe is closed: it will collapse back on itself. If matter-density is below that number, expansion will continue to dominate gravitation—the universe is "open"; it will expand indefinitely. However, if matter-density is precisely at the critical value, the forces of expansion and contraction will balance each other and the universe is "flat." It will remain balanced at the razor's edge between the opposing forces of expansion and contraction.

Recent findings disclose aspects of the universe that are unexpected, if not entirely anomalous. In light of the standard model, for example, the "Boomerang" (Balloon Observations of Millimetric Extragalactic Radiation and Geophysics) project's observations of the microwave background in 1999—observations that covered only 2.5 percent of the sky but that achieved a resolution thirty-five times higher than that of COBE—are truly surprising: they indicate that the universe is precisely flat. This finding was impressively confirmed by a number of increasing sophisticated observations: by MAXIMA (Millimeter Anisotropy Experiment Imaging Array) as well as by DASI (Degree Angular Scale Interferometer, based on a microwave telescope at the South Pole), and most recently by WMAP (Wilkinson Microwave Anisotropy Probe, a satellite that has been orbiting the Earth since June 30, 2001 and recording cosmic radiation from a point on the far side of the moon). It is now beyond reasonable doubt that the Big Bang was fine-tuned to the staggering precision of one part in 10^{50}! A deviation even of that minute order would have produced an infinitely expanding or a finite recollapsing universe.

Not only is the matter-density of the universe precisely tuned for balance between expansion and contraction; the universe's forces are also precisely tuned to the parameters of its matter particles. As Arthur Eddington and Paul Dirac already observed, the ratio of the electric force to the gravitational force is approximately 10^{40}, while the ratio of the observable size of the universe to the size of the electron is likewise around 10^{40}. This is strange, because the former ratio should be unchanging (both forces are believed to be constant), whereas the latter should be changing (since the universe is expanding). If the agreement of these ratios, the one variable the other not, is more than a temporary coincidence, as Dirac suggested in his "large number hypothesis," the force of gravitation is not constant over time. Moreover when Einstein's mass-energy relation is applied, the size of the electron ($ro = 6 \cdot 10^{-15}$ meters) turns out to be a consequence of the number of electrons in the visible universe (this is Eddington's number, approximately 2×10^{79} in the Hubble universe of $R = 10^{26}$ meters).

Menas Kafatos and his collaborators showed a relationship between the masses of the total number of particles in the universe to the gravitational constant, the charge of the electron, Planck's constant, and the speed of light (Kafatos 1989, 1990, 1999). Scale-invariant relationships appear—for example, all lengths turn out to be proportional to the scale of the universe. This suggests a staggeringly high level of coherence throughout the cosmos—according to Kafatos et al. the entire universe is nonlocal.

The coherence of the universe is also manifest in the fine-tuning of its basic parameters. The universal forces and constants are precisely tuned to the evolution of complex systems, including those associated with life. A minute difference in the strength of the electromagnetic field relative to the gravitational field would have prevented the evolution of systems of higher complexity since hot and stable stars such as the Sun would not have come about. If the difference between the mass of the neutron and the proton would not be precisely twice the mass of the electron, no substantial chemical reactions could take place. Similarly, if the electric charge of electrons and protons did not balance precisely, all configurations of matter would be unstable and the universe would consist of nothing more than radiation and a relatively uniform mixture of gases.

However, in this universe the gravity constant ($G = 6.67 \times 10^{-8}$) is precisely such that stars can form and shine long enough to allow the evolution of complex galactic structures in space, as well as of complex microstructures on the surface of planets associated with hot and stable stars. If G would be smaller, particles would not compress sufficiently to achieve the temperature and the density needed to ignite hydrogen: stars would have remained in a gaseous state. If on the other hand G were larger, stars would have formed but would burn faster and endure for a shorter time, making it unlikely that complex structures could evolve on the planets surrounding them. Likewise, if the Planck constant ($h = 6.63 \times 10^{-27}$ *erg*) would be even minutely other than it is, carbon-producing nuclear reactions could not occur in stars—and consequently complex structures based on carbon-bonding could not arise on otherwise suitable planetary surfaces. Given the actual value of G and h, and of an entire array of other universal constants (including the velocity of light, the size and mass of the electron, and the relationships between the size of the proton and the nucleus), the universe could evolve to the level of complexity we now observe (Barrow & Tipler 1986).

An additional feature of the coherence of the cosmos comes to light in the uniformity of the cosmic background radiation as well as of the galactic macrostructures. The microwave background radiation, emitted when the universe was about a hundred thousand years old, is known to be isotropic. But at the time the radiation was emitted the opposite sides of the expanding universe were already ten million light years apart. Light could only have traveled a hundred thousand of these light years—yet the background radiation (at 2.73 degrees on the Kelvin scale) is uniform throughout the presently observed universe. Moreover distant galaxies and macrostructures evolve in a uniform manner although they are not connected by physical signals, and have not been so connected since the first few microseconds in the life of the universe. If a galaxy ten billion light years from Earth in one direction exhibits structures analogous to a galaxy the same distance away in the opposite direction, then structures that are twenty billion light years from each other are structurally uniform. This cannot be ascribed to physical factors, since according to general relativity the highest rate at which signals can propagate in space-time is the speed of light, and light could reach across the ten-billion light-year distance from Earth to each of the galaxies (hence we can observe them), but it cannot reach from one of these galaxies to the other.

A sophisticated mathematical account of this "horizon problem" is furnished by the theory of cosmic inflation originally advanced by Alan H. Guth (Guth 1997). According to the cosmic inflation theory also elaborated by Andrei Linde, at the initial Planck-time of 10^{-33} seconds the cosmos expanded at a rate faster than light. This did not violate general relativity, since it was not matter that moved at these velocities, but space itself—matter (the particles that were the first to be synthesized) stood still relative to space. During inflation all parts of the universe were in immediate contact, sharing the same density and temperature. Subsequently some parts of the expanding universe fell out of contact with each other and evolved on their own. Even if light did not catch up with the circumference of the expanding universe (because the universe's circumference became larger than the distance light could have traveled during the corresponding time), all of its structures could evolve uniformly: they were connected during inflation.

Whether or not coherence in the current universe is adequately explained by inflation theory is as yet open to question. Cyclic models of the universe can explain all the facts accounted for by inflation theory based on one period of acceleration per cycle rather than a superfast acceleration followed by the relatively moderate acceleration of the Robertson-Walker universe. Moreover, we shall argue, cyclic models of the universe can be developed to offer an explanation of the observed fine-tuning of the universe's physical constants, whereas inflation theory cannot explain why the universe that arose in the wake of the Big Bang is such that it could produce complex structures, including the self-maintaining structures associated with life.

The puzzle is the selection of the vacuum fluctuations preceding the Big Bang. This is not likely to have been a random selection, since the fluctuations came in specific varieties, a small subset of all the varieties that were theoretically possible. The statistical probability that the varieties that had actually occurred would have come about purely by accident is negligibly small. According to calculations by Roger Penrose, the probability of hitting on a universe such as ours by randomly sifting through the alternative possibilities is of the order of one in $10^{10^{123}}$.

But perhaps our remarkably coherent universe did not arise in a randomly unordered vacuum pre-space, but in a vacuum ordered by prior cosmic history. The history of the cosmos may extend beyond the Big Bang: a growing number of investigators entertain the possibility that this universe arose in the context of a preexisting metauniverse or *metaverse* (Rees 1997, Steinhardt & Turok 2002). In part 2 we shall look at this scenario in more detail, since it may offer a logical explanation of the large-scale coherence of the universe we now observe.

Chapter 2

♦

Coherence in the Living World

We now set forth our inquiry into the foundations of a transdisciplinary unified theory by considering the kind of coherence exhibited in the domains of life. Nonlocal coherence at the mesoscale between the microscale of the quantum and the macroscale of the universe is just as remarkable as coherence in the world of the quantum and in the universe at large. At the mesoscale of life the anomalous form of coherence is manifested in the quasi-instant correlation of the organism's parts and components, entailing further correlations and hence coherence also between the organism and its external milieu.

Already half a century ago Erwin Schrödinger suggested that in regard to the living organism we must replace the concept of mechanical order with the notion of *dynamic* order. Dynamic order is not an order based on chance encounters among mechanically related parts; it cannot arise by interaction based on random collisions among individual molecules. There must be system-wide correlations that involve all the parts, even those that are distant from one another. Rare molecules, for example, though seldom contiguous, need to find each other throughout the organism. There would not be sufficient time for this to occur by a random process of mixing; the molecules must locate and respond to each other specifically, whether they are neighboring or distant.

Dynamic order in the organism is an order where the components are organized by a system-wide correlation that replaces randomness with a stable and dependable pattern. It is thanks to this correlation that organisms can maintain themselves in the inherently unstable, and physically entirely improbable regime far from thermodynamic equilibrium—the dynamic regime required to store and mobilize the energy for life's irreversible processes.

Forms of Intra-Organic Coherence

Living organisms are complex carbon-based thermodynamic systems operating in a water-based medium. They maintain themselves within a flow of energy from the environment, compensating for the degradation of free energy due to irreversible processes. Unless they constantly import the energy required to maintain their structure and function, their specific entropy increases, bringing them closer to the inert state of thermodynamic equilibrium. Life is physically possible only through the efficient storing and mobilization of free energy far from thermodynamic equilibrium.

Mae-Wan Ho suggests that the organism maintains itself in its inherently improbable state through the superposition of two basic processes: a nondissipative cyclic process, for which the net entropy balances out to zero ($DS = 0$), and a dissipative, irreversible process for which entropy production is greater than zero ($DS > 0$) (Ho 1993, 1994, 1996). The cyclic, nondissipative loop embraces almost all living components of the organism due to the ubiquity of the cycles that constitute it. Its coupling with the irreversible energy throughput loop frees the living organism from immediate thermodynamic constraints, as described both in the first law (the law of energy conservation) and in the second law (the law of irreversible energy degradation in closed systems).

Within the cyclic nondissipative loop the efficient mobilization of stored energy permits a high level of coherence among the components. Processes occurring in one part of the organism affect quasi-instantly all other parts, and all processes are sensitively coupled to respond to conditions in the environment. As a consequence the living state is dynamic and fluid. Its processes engage all levels simultaneously, from the microscopic to the molecular and macroscopic. Adjustments, responses and changes required for the maintenance of the whole organism propagate in all directions at the same time, and they are sensitively tuned to the organism's environment. Signals even below the threshold of biochemical sensitivity can produce macroscopic effects. For example, the eye responds even to a single photon, and the ear picks up sounds below the level of thermal noise. Extremely low-frequency electromagnetic radiation also produces a response, since a large number of molecules are coherently linked among themselves. This kind of coherence requires the instant mobilization of energy by closely coupled processes throughout the system.

The coherence of the organism confirms Hans Fröhlich's original discovery that all parts of the living matrix create fields that infuse the organism and radiate into the environment at different frequencies, including the frequency of visible light—subsequently confirmed by Fritz-Albert Popp as the emission of "biophotons" (Bischof 2002). The specific resonance frequency of each molecule, cell, tissue, and organ coordinates its behavior with all others, creating long-range phase correlations similar to those that occur in superfluidity and superconductivity.

The discovery that the organism exhibits processes typical of superfluidity was confirmed by the finding that living tissue constitutes a Bose-Einstein condensate. Originally postulated in 1924, the existence of such condensates has been experimentally verified by Eric A. Cornell, Wolfgang Ketterle, and Carl E. Wieman in 1995, in experiments for which they received the 2001 Nobel Prize in physics. The experiments show that supercooled aggregates of matter—in the experiments ribidium or sodium atoms—behave as nonlocal waves, penetrating throughout the condensate and forming interference patterns (other elements, including hydrogen, have been found to display Bose-Einstein condensation as well). This process also occurs in living tissue and accounts for certain aspects of the organism's nonlocal coherence.

Nonlocal coherence in living systems suggests that the organism constitutes a macroscopic quantum system. Molecular assemblies, whether neighboring or distant, resonate in phase: the same wavefunction applies to them. The phase relations determine whether the force carried in the assemblies is attractive or repulsive. Faster and slower reactions accommodate themselves within an overall process where the component wavefunctions coincide. Thereby long-range correlations come about that are nonlinear, quasi-instant, heterogeneous, as well as multidimensional.

Quantum systems have remarkable and highly pertinent properties. In such systems a non-substantial mode of information transfer takes place, using entangled states as channels of quantum information. (In this context entangled states are states of distant and classically non-interactive components of the system that cannot be accounted for by the individual properties of the components.) In the organism molecular reactions at different space-time points carry out individual functions, but the co-ordination of the functions is ensured by quantum coherence.

A macroscopic quantum system differs in essential respects from a classical system. A quantum system does not allow a precise determination of position and momentum and other non-communicating variables, there is no continuous change in energy, entropy and information, identical parts do not have a separate identity, and there is no individual determination of the diverse attributes of the system. The quantum system does allow, on the other hand, nonclassical processes such as tunneling through a potential barrier, interference among all possible histories, sensitivity to electromagnetic potentials, entangled states, and teleportation. According to R. P. Bajpai these processes confer remarkable capabilities on the living system, such as perfectly secure communication, algorithmic searching (Grover's algorithm), rapid factoring of large numbers (Shor's algorithm), efficient simulation of other quantum systems, almost loss-less channels of information transfer, and signal detection below noise level (Bajpai 2002). These processes are currently investigated in the new field of quantum biology and, as Bajpai notes, are highly encouraging for our understanding of the phenomenon of life. An integrated and unified vision of living and non-living systems seems to be emerging: a quantum vision of life and world.

The concept of the living organism as a macroscopic quantum system places in question the dominant paradigm for the investigation of the living state. The essential features of this state appear to be not molecular and biochemical, but quantum-level and quantum-physical.

Independent evidence speaks to the assumption that molecular determinism (the view that living processes can be adequately accounted for in reference to molecular interactions) is obsolete; even genetic determinism (the claim that the set of genes in the genome contains a complete set of instructions for building and operating the organism) may be open to question. While it is reasonably certain that by means of complementary copies of messenger-RNA genes determine the amino acid sequence of protein molecules, it is not clear that this primary structure would determine all the higher-order structures and functions of the organism. Almost all cells of the organism are genetically equivalent; they become differentiated in the embryo because in different cells different sets of genes are activated and suppressed. Some cells constitute building blocks, others play the role of enzymes, another class participates in cell-signaling, and some (the motor proteins) transform chemical energy into mechanical energy. According to genetic determinism, the process of differentiation is entirely genetic: other genes are responsible for it. It is possible, however—and in the quantum-system conception of the organism entirely plausible—that some basic developmental processes are either outside of genetic control, or are only indirectly affected by genes. Lev Beloussov suggests that the truth may be the reverse of genetic determinism: genes themselves could be obedient servants fulfilling powerful commands by the rest of the organism (Beloussov 2002).

Genetic determinism encounters further empirical paradoxes. These include the C-value paradox (where C stands for complexity and C-value denotes the size of the organism's haploid set of chromosomes, that is, the size of its DNA sequence), and the gene-number paradox (also known as the paradox of gene redundancy).

The C-value is paradoxical inasmuch as it fails to meet expectations. One would assume that if the genome possesses a reasonably complete description of the organism, the complexity of the phenome and the complexity of the genome are proportional: more complex organisms have more complex genetic structures. This, however, is not the case. A simple amoeba has two hundred times the number of DNA per cell of Homo sapiens. Even when organisms are phylogenetically related, they sometimes have radically different genomes. The genome size of closely related rodents often varies by a factor of two, and the genome of the housefly is five times larger than the genome of the fruit fly. At the same time some phylogenetically distant organisms have similarly complex genetic structure. It appears that the complexity of the phenome is not necessarily reflected in the complexity of the genome. This is a paradox for the standard view that instructions coded in the genome determine the structure and function of the phenome.

The gene-number paradox derives in turn from the finding that the number of genes found by molecular methods is always larger than the number of genes that are accessible to genetic analysis. There is a significant excess of genes beyond those for which a phenotypic function can be established. A wide variety of genes can be mutated without a deleterious effect on the organism; many can be mutated without any effect at all. Genes are often copied in the genome with minor modification; it appears that all gene-copies would have to mutate in order to perturb organic function. These findings are anomalous for genetic determinism. Although a genetic basis exists for many organic features and functions, the highly coherent structure and function of the living organism cannot be adequately explained solely in reference to molecular interaction and genetic instruction.

Aspects of Transorganic Coherence

The living organism is not only internally coherent; it also exhibits a high level of coherence with its milieu. Transorganic coherence has its roots within the organism. It appears that the connectivity ensured by the coupling of energy flows throughout the organism also links the genome with the phenome, and through the phenome links the organism with its pertinent milieu. Evidence is becoming available that the genome is "fluid"— a flexible part of the energetically integrated organism.

Empirical evidence comes from evolutionary as well as from experimental biology. The evidence of evolutionary biology is indirect: it concerns the statistics of probability. The oldest rocks date from about 4 billion years before our time, while the earliest and already highly complex forms of life (blue-green algae and bacteria) are over 3.5 billion years old. Yet the assembly even of a primitive prokaryote involves building a double helix of DNA consisting of some 100,000 nucleotides, with each nucleotide containing an exact arrangement of thirty to fifty atoms, together with a bilayered skin and the proteins that enable the cell to take in food. This construction requires an entire series of reactions, finely coordinated with each other. If living species had relied on chance variation in the genome alone, this level of complexity is not likely to have emerged within the relatively brief period of five hundred million years.

The evolution of species through the random mutation of the genome in living populations encounters additional puzzles. They are rooted in the recognition that it is not enough for mutations to produce one or a few positive changes in a species; if they are to be viable, they must produce the full set. The evolution of feathers, for example, does not produce a reptile that can fly: radical changes in musculature and bone structure are also required, along with a faster metabolism to power sustained flight. Each innovation by itself is not likely to offer evolutionary advantage; on the contrary, there is a significant probability that it will make an organism less fit than the standard form from which it departed. And if so, it would soon be eliminated by natural selection.

The statistical improbability of random mutations to produce viable mutants does not mesh with the tenets of Darwinian and neo-Darwinian biology. The "synthetic theory" maintains that random processes of genetic mutation exposed to natural selection evolve one species into another by producing new genes and new developmental genetic pathways, coding new and more viable organic structures, body parts, and organs. Yet mutation in the genome is by no means a simple process, at least nine varieties of genetic rearrangements are known (transposition, gene duplication, exon shuffling, point mutation, chromosomal rearrangement, recombination, crossing-over, pelitropic mutation, and polyploidy), and many of them are interrelated. For example, transposition, exon shuffling, as well as gene duplication can lead to the duplication of entire genetic sequences. The assumption that these mechanisms, singly or in combination, produce new species from old by chance variation faces the problem of creating complexity within finite times. The "search-space" of possible genetic rearrangements within a genome is so enormous that random processes are likely to have taken far longer to produce viable species than the empirically known timeframes.

As early as 1937, Theodosius Dobzhansky noted that the sudden origin of a new species by gene mutation might be an impossibility in practice (Dobzhansky 1982). With the known mutation rates, he wrote, the probability of an event that would catapult a new species into being through simultaneous changes in many gene loci and various chromosomal reconstructions is negligible. Dobzhansky assumed that species formation is a slow and gradual process, occurring on a "quasi-geological scale." However, as of the 1970s Stephen Jay Gould and Niles Eldredge have contested the quasi-geological timescale of evolution (Eldredge 1985, Eldredge and Gould 1972, Gould 1983, 1991, Gould and Eldredge 1977). They emphasized that most peripherally isolated populations are relatively small, and undergo characteristic changes at a rate that translates into geological time as an instant. Speciation seems to occur on the periphery of a population in a timespan often of no more than five to ten thousand years, interspersing long periods of relative stability. The species that emerge in the process of "punctuated equilibrium" change only through random genetic drift.

Punctuated equilibrium theory's reduced timeframe makes it even more questionable that a series of random genetic rearrangements can produce viable genetic pathways, and organs and body parts coded by new genes. While random mutations can produce variants of every gene, and can also produce defective genes, they are not likely to evolve one organism into a distinctly different organism with a new genetic and body structure within the known dimensions of evolutionary time. It is difficult to see, for example, how a random rearrangement of the genome could produce the blood coagulation system of a mammal with its many unique genes, or its placenta with its similarly unique genes and biochemical functions, from invertebrate organisms that entirely lack these genes. Even if, as some geneticists assume, genes "talk" to each other by way of introns in coordinating the expression of multigene systems, and even if the intervening sequence of one gene serves as the coding region for other genes, the search-space of the rearrangement processes remains astronomically large, and the process of creating viable mutants remains subject to numerous pitfalls: as Michael J. Behe pointed out, every step in the process of mutating a preexisting "irreducibly complex" system into another such system must include all the parts of the system in a functional relationship. Living systems are so constituted that missing but a single step in the process of transformation leads to a dead-end (Behe 1998). Consequently a macromutation that involves the reorganization of the entire chromosomal material is not likely to occur except as an isolated instance without evolutionary significance.

Complex organisms are not likely to come about in nature through random processes. According to Fred Hoyle the probability of life evolving through random genetic variation is about the same as the probability of a hurricane blowing through a scrap yard assembling a working airplane (Hoyle 1983).

In addition to the negative probability-based evidence furnished by the fossil record, there is positive experimental evidence for effective linkages between the genome and the phenome. The most obvious of these linkages is physiological: it is conveyed by mechanical force. A. Maniotis and collaborators described an experimental situation where mechanical forces impressed on the external cellular membrane were transmitted via the cytoskeletonal-complex of inner cellular microtubules to the nucleus—this produced quasi-instantaneous rearrangements within the nucleus (Maniotis et al. 1997). The transmission mechanism is becoming elucidated: the endoplasmic reticulum, an inner vesicular membrane, is continuous with the nuclear and the outer-cellular membrane and in several regions it is connected to the cytoskeleton. The latter in turn is continuous with other cellular membranes and organelles within the membrane system.

Michael M. Lieber's experiments indicate that mechanical force acting on the outer membrane is but one variety of interactions that results in a mutation in the genome. It appears that any stressful force from the environment, whether mechanical or other, conveyed to the genome nonuniformly via the cell's external and internal membrane system triggers a global hypermutation (Lieber 1998a, 1998b). Mutations that are seemingly not adaptive in themselves occur inasmuch as they are necessary for the occurrence of those that are. The genome responds mutagenetically as a global dynamic whole.

Through its force connections to the environment, the organism is driven to complete within itself every stress-factor to which it is exposed. The stress-factor can be mechanical, as well as electromagnetic or even nuclear, acting temporarily in the long-range. The completion process involves the instantaneous nonlocal generation of the complete force-configuration. In Lieber's view, the drive toward completion may involve an interrelationship among various forces and may reflect and utilize a deeper connection among them.

The mutagenetic stabilization-response of the organism to external stress is evident when electromagnetic or radioactive fields irradiate the organism: this appears to have direct effect on the structure of the genes and can produce heritable variations. Experiments in Japan and in the United States have shown that rats develop diabetes when a drug administered in the laboratory damages the insulin-producing cells of their pancreas, and that the diabetic rats produce offspring in which diabetes arises spontaneously. It appears that the alteration of the rats' somatic cells produces corresponding alterations in the DNA of their germline.

Even more striking are experiments in which particular genes of a strain of bacteria are rendered defective. It turns out that some bacteria mutate back precisely those genes that were made inoperative. Exposure to chemicals produces adaptive mutations as well. It is known that when plants and insects are subjected to toxic substances, they occasionally mutate their gene pool in such a way as to detoxify the toxins and create resistance to them.

The tenet regarding the isolation of the genome is falsified both negatively, through statistical probability, and positively, by way of laboratory experiments. Adaptively responsive mutagenesis exists; genetic mutations are not isolated events, and although some mutations may be random, they are not always and necessarily so. As regards their evolutionary significance, mutations may be considered dynamic responses on the part of a species to the physical, chemical, climatic, and other changes individuals in successive generations experience in their milieus. Genome and phenome form an integrated system of functionally autonomous parts, correlated so as to survive, reproduce, and evolve as a coherent whole.

Chapter 3

✦

Coherence in the Sphere of Mind

Coherence in the living world stands to reason: if they are to survive and evolve in the inherently improbable dynamic regime far from thermodynamic equilibrium, organisms need to be highly coherent both in regard to their own structure, and in relation to their environment. But there is a surprising and as yet anomalous level of coherence coming to light in the sphere of the human mind as well. The coherence that comes to light is not just the coherence of the individual brain and mind, the expression of the integrity of the individual organism, but a potential for coherence among the brains and minds of different people: *transpersonal* coherence.

Transpersonal coherence is an important pillar of transdisciplinary unification in the here envisaged integral science. It is discovered in some of the latest branches of consciousness research. Research on consciousness is enjoying a renaissance: it is pursued in a variety of schools with a diversity of approaches. Developmental psychologists investigate the flow of living experience as an unfolding process with a specific architecture at various stages of growth; social psychologists ascribe lived experience to the networks of cultural meaning that emerge in society. Neuropsychologists maintain that consciousness is produced in the brain through neural networks, neurotransmitters, and other cerebral information-processing mechanisms; and cognitive scientists, while sharing this assumption, create functional schemata of the workings of the conscious brain in terms of organized integrated networks. In psychoneuroimmunology, psychosomatic medicine, and other forms of biofeedback research attention centers on the connection between consciousness and bodily processes, while investigators of altered states of consciousness analyze the effects of dreams, psychedelic substances, trance, and meditative states on the assumption that these disclose important and otherwise hidden aspects of consciousness. Quantum brain theory, the newest branch of consciousness research, builds models of the interaction of consciousness with the physical world, using concepts such as nonlocality, entanglement, phase-relations, hyperspace, and wavefunction.

Despite great diversity in method and scope in contemporary consciousness research, most investigators agree that the phenomenon of mind is present both in the individual, and in the sphere of information and communication created by intercommunicating individuals. Mind and consciousness are both personal, and transpersonal.

Gestalt psychologists have been pointing to the coherence of the individual mind for decades. Confronted with what the Greeks called the "booming buzzing confusion" of sensory experience, the human mind selects compatible elements and organizes them into meaningful and irreducible wholes, *Gestalts*. These are not just an assembly of their parts: they have their own properties that are conserved even when all their parts undergo transformation. For example, a melody in music is perceived as a whole and not as a mere collection of sounds. And it is perceived as one and the same whole even when the sounds of which it consists are transposed into a different key (so that all their frequencies are different) and even when they are produced by different instruments (in which case all their timbres are different).

The wholes of Gestalt psychology are elements of personal experience, the result of selection and organization by the individual mind. By contrast the kind of coherence that now emerges in consciousness research is coherence among the minds of different individuals, that is, transpersonal coherence.

Traditionally, transpersonal coherence has been noted by anthropologists who discovered that members of so-called primitive tribes can occasionally remain in extrasensory contact with one another. Australian aborigines, for example, seem capable of receiving information on traumatic events concerning their tribe or family even when they roam beyond the range of seeing and hearing (Elkin 1942, Morgan 1991). Today, however, transpersonal coherence is coming to light in the laboratory, as experimental parapsychologists investigate thought and image transference and similar "psi-phenomena" (Braud 1992). Such phenomena have been known to occur among identical twins as well as between mothers and sons and other persons with close emotional ties. It now appears that the phenomenon is more widespread than it was thought. Controlled tests show that many if not all individuals possess the corresponding capacity.

Transpersonal Coherence: The Mental Aspect

Russell Targ and Harold F. Puthoff pioneered controlled experiments on thought and image transference in the early 1970s (Puthoff & Targ 1976, Targ & Harary 1984, Targ & Puthoff 1974). They placed one subject, the "receiver," in a sealed, opaque, and electrically shielded chamber, and placed another, the "sender," in a remote location where she was subjected to bright flashes of light at regular intervals. The brain-wave patterns of both sender and receiver were registered on electroencephalograph (EEG) machines. As expected, the sender exhibited the rhythmic brain waves that normally accompany exposure to bright flashes of light. However, after a brief interval the receiver also began to produce the same patterns, although she was not being directly exposed to the flashes and was not receiving ordinary sense-perceivable signals from the sender.

In remote viewing experiments distances ranging from a few hundred yards to several miles separated sender and receiver. At a randomly chosen site the sender acted as a "beacon," and the receiver was asked to pick up what the sender saw. To document their impressions receivers gave verbal descriptions, sometimes accompanied by sketches. Independent judges found that the descriptions of the sketches matched the characteristics of the site actually seen by the sender on the average 66 per cent of the time.

Remote viewing experiments reported from a variety of sources and laboratories reported a success rate around 50 percent, considerably above random probability. The most successful viewers appeared to be those who were relaxed, attentive, and meditative. They reported receiving a preliminary impression as a gentle and fleeting form, which gradually evolved into an integrated image. They experienced the image as a surprise, both because it was clear in itself, and because it was clearly elsewhere.

Jacobo Grinberg-Zylberbaum at the National University of Mexico performed more than fifty experiments on transpersonal transference among highly shielded subjects over a period of five years (Grinberg-Zylberbaum et al. 1993). He paired his subjects inside soundproof and electromagnetic radiation-proof "Faraday cages" and asked them to meditate together for twenty minutes. Then he placed them in separate Faraday cages where one subject was stimulated and the other not. The stimulated subject received stimuli at random intervals in such a way that neither she nor the experimenter knew when they were applied. The subjects who were not stimulated remained relaxed, with eyes closed, instructed to feel the presence of the partner without knowing anything about her stimulation. In general, a series of one hundred stimuli were applied (flashes of light, sounds, or short, intense, but not painful electric shocks to the index and ring fingers of the right hand). The EEG records of both subjects were then synchronized and examined for "normal" potentials evoked in the stimulated subject and "transferred" potentials in the nonstimulated person. Transferred potentials were not found in control situations without a stimulated subject, when a screen prevented the stimulated subject from perceiving the stimuli (e.g., light flashes), or when the paired subjects did not previously interact. But during experimental situations with stimulated subjects and with prior contact among them the potentials appeared consistently in about twenty-five percent of the cases.

In a limited way, Grinberg-Zylberbaum could also replicate his results. When one individual exhibited the transferred potentials in one experiment, he or she usually exhibited them in subsequent experiments as well. The results did not depend on spatial separation between senders and receivers—the potentials appeared no matter how far or how near they were to each other. Experiments replicating these results have been carried out subsequently at Bastyr University as well as the University of Washington (Richards & Standish 2000, Thaheld 2001).

Cleve Backster performed prima facie different but on deeper analysis closely related experiments in the United States. Backster, an expert on the use of the polygraph (lie detector), first attracted attention in the 1960s when he applied the electrodes of his instrument to the leaf of a plant in his office and found that the instrument registered the kind of reactions by the plant that he himself has undergone (Backster 1968, 1975). His claim of a "primary perception" on the part of the plant was controversial, especially since he obtained the analogous response when he detached the leaf he tested from the plant, and even when the detached leaf was trimmed to electrode size or shredded and redistributed between the electrode surfaces.

Backster's subsequent experiments involved in vitro cells detached from human subjects. Oral leukocytes (white cells in the mouth) were taken from test subjects and tested at distances ranging from 5 yards to over 8 miles (12 km). The cells were removed using the standard Klinkhamer procedure, and the solution was centrifuged to transfer the white cell yield to a 1 ml. culture tube in preparation for electroding. Two flexible wire leads were attached to the top of gold electrodes, and the net electrical potential activity of the cells was monitored, using an EEG-type instrumentation. The signals were fed to a device where a chart drive-unit provided continuous recording with permanent ink tracings. The display provided a graphic read-out of net electrical potential changes in the leukocytes.

The test subjects were stimulated with visual materials designed to evoke an emotional response, and the variations in the electrical potential of the leukocytes were correlated with the responses of the subject in regard to time, amplitude, and duration. In one experiment a male subject age 25 was handed an issue of *Playboy* magazine and perused its pages during the test. When he came to the centerfold photo, a nude picture of actress Bo Derek, he manifested an emotional response that lasted during the whole time he was focused on the picture. When he closed the magazine his responses returned to average values. However, when he decided to reach for the magazine again, a second reaction of the same kind took place. After the test the subject confirmed to have had an emotional arousal at these times. The tracings of the electrical potentials of his detached leukocytes, located at a distance of 5 yards, showed a strong correlation, with sudden fluctuations at the precise time of his reactions.

In another test the subject was a professional animal trainer who trained marine mammals for aquaria. He was comfortable working with killer whales and leopard seals and other large predatory marine mammals, without feeling fear. During the test he discussed working with marine mammals, and during this time no reactions were registered by the intrument attached to his leukocytes 5 yards distant. However, when an image of tigers and leopards in close proximity to him was evoked, the tracings showed wide fluctuations. A similar response was observed in the case of a retired U.S. Navy gunner stationed at Pearl Harbor during the Japanese attack. When watching a TV program entitled "The World At War," he did not react to the downing of enemy aircraft by naval gunfire except when it occurred immediately following a facial close-up of a naval gunner in action. At that point—when he apparently projected his own wartime experiences into the scene—his cells, located at a distance of 8 miles, showed an immediate reaction. He subsequently confirmed emotional arousal at that point.

In these and numerous other cases the reactions produced by distant in vitro cells were exactly those one would expect to find when the sensors are attached directly to the body of the subject. Backster could not furnish a realistic explanation why these reactions should occur in cells removed from the subjects and located at various distances them. These, he noted, are instances of biocommunication involving a signal about which there is inadequate information (Backster 1985).

An experiment in Italy carried out by physician Nitamo Montecucco measured the degree of harmonization of the EEG-waves of single subjects, as well as of groups of subjects (Montecucco 2000). In ordinary waking consciousness the two hemispheres of the brain exhibit uncoordinated, randomly diverging EEG patterns. When the subject enters a meditative state, these patterns tend to become synchronized, and in deep meditation the two hemispheres fall into a nearly identical pattern. Montecucco's EEG record shows that in deep meditation not only the left and right brains of one and the same subject, but also the left and right brains of an entire group of subjects achieve a high level of synchronization. In one experiment with twelve individuals a 98 percent synchronization obtained among eleven members of the group, in the complete absence of sensory contact between them (Montecucco 2000).

Information entering the mind beyond the range of the senses has been researched by transpersonal psychiatrist Stanislav Grof (Grof 1988, 1993, 1996, 2000). He found that in altered states of consciousness—generated either by psychotropic substances or by rhythmic breathing exercises—individuals can experience practically any aspect of the world around them without sensory contact. When experiencing the mind of other individuals, some subjects report a loosening and melting of the boundaries of the body ego and a sense of merging with another person in unity and oneness, while others achieve a sense of complete identification to the point of losing awareness of their own identity. In still deeper altered states some individuals can expand their consciousness to an extent where it encompasses the totality of life on the planet, and seems to extend outward into the cosmos.

The speculative reaches of contemporary consciousness studies are matched in a more modest yet equally remarkable way in the day-to-day experience of psychologists, psychiatrists, and psychotherapists. The pertinent experience is known as therapist-to-patient "transference," and patient-to-therapist "counter-transference." In these processes the subconscious, and occasionally the conscious, mind of the patient is infused with the feelings, images, and intuitions of the therapist—and vice versa, the mind of the therapist manifests elements directly intuited from the mind of the patient. The phenomenon is widely known and is often interpreted in reference to an interpersonal field (Schwartz-Salant 1988, Mansfield & Spiegelman 1995). Robert Langs postulated a "bipersonal field" and Michael Conforti added that through this field patient and therapist tune into a more general field whose origin is archetypal and whose characteristics contain elements from their respective dynamics (Conforti 1999).

Transpersonal contact obtains not only among the mind of individuals, but on occasion also among entire groups of people. The evidence for transgroup—or transculture—contact is circumstantial: it relies primarily on the archeological and historical record. It appears that artifacts of remarkable similarity were produced by cultures that are not likely to have been in any ordinary form of communication with each other. In widely different locations and partly at different historical times, ancient peoples developed an array of highly similar tools and buildings. Giant pyramids were built in ancient Egypt as well as in pre-Colombian America with remarkable agreement in design. Crafts, such as pottery making, took much the same form in all cultures. The Acheulean hand ax, for example, a widespread tool of the Stone Age, had a typical almond or tear-shaped design chipped into symmetry on both sides. In Europe this ax was made of flint, in the Middle East of chert, and in Africa of quartzite, shale, or diabase. Its basic form was functional, yet the agreement in the details of its execution in virtually all traditional cultures cannot be explained by the simultaneous discovery of utilitarian solutions to a shared need: trial and error is not likely to have produced such similarity of detail in so many far-flung populations.

At this writer's suggestion, University of Bologna historian Ignazio Masulli made an in-depth study of the pots, urns, and other artifacts produced by indigenous and independently evolving cultures in Europe, as well as in Egypt, Persia (now Iran), India, and China during the period from the 6th to the second millennia BC (Masulli 1997). Masulli found striking recurrences in the basic forms and designs, and he could not come up with a conventional explanation. The civilizations lived far apart in space and sometimes also in time, and did not seem to have had conventional forms of contact with each other. The phenomenon is widespread. Although each culture added its own embellishments, Aztecs and Etruscans, Zulus and Malays, classical Indians and ancient Chinese built their monuments and fashioned their tools as if following a shared pattern or archetype.

Transpersonal Coherence: The Somatic Aspect

Not only thoughts, images, and stimuli, also bodily effects can be transmitted and received across space and time. These "telesomatic" (rather than telepathic) transpersonal effects consist of physiological changes triggered in the organism of the receiver by intentional processes in the mind of the sender.

The pioneering study in the area of telesomatic effect transmission is the work of cardiologist Randolph Byrd, a former professor at the University of California at Berkeley (Byrd 1988). His ten-month computer-assisted study concerned the medical histories of patients admitted to the coronary care unit at San Francisco General Hospital. Byrd formed a group of experimenters made up of ordinary people whose only common characteristic was a habit of regular prayer in Catholic or Protestant congregations around the country. The selected people were asked to pray for the recovery of a group of 192 patients; another set of 210 patients, for whom nobody prayed in the experiment, made up the control group. Rigid criteria were used: the selection was randomized and the experiment was carried out double-blind, with neither the patients nor the nurses and doctors knowing which patients belonged to which group. The experimenters were given the names of the patients, some information about their heart condition, and were asked to pray for them every day. They were not told anything further. Since each experimenter could pray for several patients, each patient had between five and seven people praying for him or her.

In terms of the statistics of probability, the results were highly significant. The prayed-for group was five times less likely than the control group to require antibiotics (3 versus 16 patients); it was three times less likely to develop pulmonary edema (6 compared to 18 patients); none in the prayed-for group required endotracheal intubation (while 12 patients in the control group did); and fewer patients died in the former than in the latter group (though this particular result was statistically not significant). It did not matter how close or far the patients were to those who prayed for them, nor what type of praying was practiced—only the fact of concentrated and repeated prayer seemed to have counted, without regard to whom the prayer was addressed and where it took place. A subsequent experiment regarding the effect of remote prayer carried out under still more stringent conditions by a different team of investigators showed similarly significant results (Harris et al., 1999).

Physiological effects can also be transmitted in the form anthropologists call "sympathetic magic." Shamans, witch doctors, and those who practice such magic—voodoo, for example—act not on the person they target, but on an effigy of that person, such as a doll. The practice is widespread among traditional people; the rituals of Native Americans make use of it as well. In his famous study *The Golden Bough*, Sir James Frazer noted that Native American shamans would draw the figure of a person in sand, ashes, or clay, and then prick it with a sharp stick or do it some other injury. The corresponding injury was believed to be inflicted on the person the figure represented. Observers found that the targeted person often fell ill, became lethargic, and would sometimes die (Frazer 1899). Dean Radin and his collaborators at the University of Nevada decided to test the positive variant of this effect under controlled laboratory conditions (Radin 1997).

In Radin's experiments the subjects created a small doll in their own image, and provided various objects (pictures, jewelry, an autobiography, and personally meaningful tokens) to "represent" them. They also gave a list of what makes them feel nurtured and comfortable. These and the accompanying information were used by the "healer" (who functioned analogously to the "sender" in thought- and image-transfer experiments) to create a sympathetic connection to the "patient." The latter was wired up to monitor the activity of his or her autonomous nervous system—electrodermal activity, heart rate, and blood pulse volume—while the healer was in an acoustically and electromagnetically shielded room in an adjacent building. The healer placed the doll and other small objects on the table in front of him and concentrated on them while sending randomly sequenced "nurturing" (active healing) and "rest" messages.

In the experiments the electrodermal activity of the patients, together with their heart rate, were significantly different during the active nurturing periods than during the rest periods, while blood pulse volume was significant for a few seconds during the nurturing period. Both heart rate and blood flow indicated a "relaxation response"—which made sense since the healer was attempting to "nurture" the subject via the doll. On the other hand, a higher rate of electrodermal activity showed that the patients' autonomous nervous systems was becoming aroused. Why this should be so was puzzling until the experimenters realized that the healers nurtured the patients by rubbing the shoulders of the dolls that represented them, or stroked their hair and face. This seems to have had the effect of a "remote massage" on the patients! Radin concluded that the local actions and thoughts of the healer are mimicked in the remote patient almost as if healer and patient were next to each other.

Radin's findings have been corroborated by William G. Braud and Marilyn Schlitz in hundreds of experiments carried out over more than a decade (Braud 1992, Braud & Schlitz 1983). The experiments tested the impact of the mental imagery of senders on the physiology of receivers. The effects proved similar to those produced by the subjects' own mental processes on their own body. "Telesomatic" action by a distant person is nearly as effective as "psychosomatic" influence by the subjects themselves.

In the newest form of medical practice physician Larry Dossey calls "Era III nonlocal medicine," telesomatic effects are used for healing (Dossey 1989, 1992, 1993). A sensitive is asked to concentrate on a given patient from a remote location. As shown in the practice of numerous healers, it is enough to give the name and date of birth of the patient. Neurosurgeon Norman Shealy, for example, telephoned this information from his office in Missouri to clairvoyant diagnostician Carolyn Myss in New Hampshire, and found that in the first one hundred cases the latter's diagnosis was 93 percent correct.

In the United Kingdom a group of accredited medical doctors practice another form of remote healing: "psionic medicine" where "psi" refers to the psi field that is said to envelope the organism (*Psionic Medicine* 2000, Reyner 2001). For accessing this field the members of the Psionic Medical Society use a sophisticated form of dowsing for diagnosis (observing the movement of a hand-held pendulum over a specially designed medical chart), and homeopathic remedies for treatment. Diagnosis is not necessarily effected on the patient directly; it can be performed remotely by means of a so-called witness, which can be any protein sample of the patient's organism such as a strand of hair or drop of blood. The witness can be analyzed repeatedly, and at any time and any distance from the patient. The information it produces is not limited to the state of health of the patient at the time the sample was taken, but reflects his or her state of health at the time of the analysis (which would not be the case if the unchanging—and progressively degenerating—cellular structure of the sample were the source of the information). Psionic diagnosis has proved correct in thousands of cases over several decades, in the practice of more than a hundred eminent physicians.

Austrian engineer Erich Koerbler developed a related technique, called "new homeopathy." As practiced in Hungary, Austria, and Germany by Koerbler's student and successor Maria Sági, it involves diagnosis with the help of a specially developed medical dowsing rod together with a vector system of geometrical forms developed by Koerbler (Sági 1998). Treatment is effected either by homeopathic remedies, or by the healer "sending" the geometric forms and other healing symbols derived inter alia from traditional Eastern practices. Sági's method of diagnosis and healing has proved to be effective in a decade-long practice, on patients both locally and in diverse parts of the world, and has been experienced by this writer.[1] The positive results of this form of healing is by no means exceptional: a systematic review of the efficacy of various kinds of distant healing in different locations showed that approximately 57 percent (13 of 23) of randomized, placebo-controlled trials had a positive treatment effect (Astin et al. 2000).

---

1. Controlled experiments with this method are reported by Dr. Sági in Appendix 2.

Chapter 4

✦

Understanding Coherence

The Elements of an Explanation

The finding of the above reviewed forms of coherence in diverse domains of nature and mind offers a basis for the transdisciplinary unification of our understanding of both cosmos and consciousness. It needs to be integrated into the conceptual structure of science, however, and this may call for the revision of some fundamental assumptions. While phenomena are always interpreted in light of some preexisting conceptions, phenomena do not necessarily fit smoothly into those conceptions. Those that refuse to fit are considered anomalous. This is true of many of the pertinent forms of coherence. Quantum-coherence through nonlocal correlation is recognized in particle physics and is beginning to be recognized in cosmological physics as well, but it is not explained by relativity theory (where the maximum speed of signal transmission is the speed of light), and quantum mechanics, while it accounts for nonlocality through mathematical formalisms, fails to produce an explanation in terms of physical processes. The Darwinian paradigm in the life sciences cannot explain the full scope of organic coherence, and the current branches of consciousness research, confronted with evidence for various forms of transpersonal coherence, are as yet at a loss to account for them.

Given this situation, scientists either produce auxiliary assumptions and hypotheses to "build in" the anomalous phenomena, or look for fundamentally new assumptions to explain them. Beyond a given point of anomaly-accumulation, the proliferation of ad hoc additions, like the "epicycles" of pre-Copernican astronomy, becomes conceptually burdensome as well as unmanageably complex. At that point the search begins for new concepts to ground a different fundamental framework—a new "paradigm." This point has now been reached in a number of avant-garde research communities. This chapter suggests the foundations of the paradigm that could ground an acceptable explanation of coherence phenomena in the various domains of its observation.

The reasoning behind the novel paradigm is encapsulated in the following propositions:

> Anomalous coherence in a system implies quasi-instant correlation among the parts and components of that system.
>
> Such correlation implies system-wide connectivity.
>
> System-wide connectivity implies in turn the presence of an interconnecting medium.
>
> In a realist perspective the interconnecting medium is a system-wide field.

We conclude that *anomalous coherence implies a system-wide field.* Let us examine this tenet in more detail.

Coherence, we have seen, is essentially a correlation phenomenon. We have said that the parts of an intrinsically correlated system are connected in such a way that what happens in and to one of its parts also happens in and to all its other parts, and hence it happens in and to the system as a whole. In consequence the parts respond to the "rest of the world" as a whole, maintain themselves as a whole, and change and evolve as a whole. This kind of correlation is not produced either by mechanical or chemical interaction among the parts: in most cases these are too slow and limited to produce the observed phenomena. The kind of correlation involved here is intrinsic and quasi-instantaneous. It is independent of the limitations of space and time, occurring over times and spaces that range from the ultrasmall Planck-times and Planck-dimensions to cosmological times and distances.

Is intrinsic quasi-instantaneous correlation purely a phenomenon, or does it have a scientifically valid explanation? The latter, we argue, is the case. There is a physical factor that can render such correlation realistically conceivable. It is best conceptualized as a *field*.

The concept of field stems from classical mechanics, where the need to link events at different points in space arose out of Newton's theory of gravitation. If one event at one point in space attracts another event at another point, there has to be some way of transmitting the effect from the first point to the second. In the eighteenth century physicists began to interpret gravitational attraction as action in a gravitational field: this field was assumed to be built by all the existing mass-points in space and to act on each mass-point at its specific spatial location. In 1849, Michael Faraday used this notion to replace direct action among electric charges and currents with electric and magnetic fields produced by all charges and currents existing at a given time, and in 1864 James Clerk Maxwell stated the electromagnetic theory of light in terms of the field in which electromagnetic waves propagate at finite velocity. Since the middle of the twentieth century, the four classical fields—the gravitational, the electromagnetic, and the strong and the weak nuclear fields—have been joined by a variety of so-called nonclassical fields, postulated in quantum field physics. In recent years these have come to be viewed as manifestations of a unified energy domain, associated with the quantum vacuum.

Fields are highly abstract theoretical entities, but they are not necessarily limited to the domain of theory. "Classical" fields, such as the gravitational, the electromagnetic, and the strong and the weak nuclear fields are said to exist independently of the theories and observations in the framework of which they are postulated. Other fields can be considered physically real as well, among them the torsion and spinor fields (but not the probability fields) postulated in quantum field theory.

Physically real fields are not themselves observable, but they produce observable effects. Anomalous coherence in nature could be the effect of such a field (or fields). If so, we should identify its (or their) nature and origins. To this end we now review the principal varieties of fields postulated at the leading edge of contemporary physical, biological and psychological research.

Physical Fields

Fields are a widely used construct in physics. Quantum theory postulates various fields, some clearly abstract (e.g., probability fields) and others of more definite reality-status. In relativistic quantum field theory (the theory that integrates special relativity, quantum mechanics, and field theory) both matter and force are defined by quantized field descriptions. But physically real fields are seldom invoked to account for quantum nonlocality. In the mainstream physics community Nils Bohr's interdiction holds sway: correlate the *observations*, and do not seek the *observables* that would underlie them. This makes the world of the quantum mathematically consistent, but realistically weird.

Weirdness extends to the latest developments in string and superstring theory. In these theoretical frameworks space-time ceases to be smooth at very small scales; it is not flat even in the absence of mass, but constitutes a turbulent "quantum foam." The infinities generated by this turbulence are eliminated in string theory by "smearing" the short-distance properties of space to smooth the quantum turbulence. To this end the elementary furnishings of the universe are seen not as point particles, but as one-dimensional vibrating filaments or "strings." To observers they appear as particles because current instrumentation cannot penetrate to the required scale, which is the Planck-length of 10^{-35} m (technology allows measurements only down to 10^{-18} m). String theorists claim that electrons, muons, and quarks, as well as the entire class of bosons (light and force particles) and of fermions ("matter" particles) are different vibrations of the string, their properties a consequence of the universe's overall geometry. Particles correspond to the lowest vibration patterns, represented as "holes" in a Calabi-Yau space. Their masses are determined by the way in which the boundaries of Calabi-Yau space-holes intersect and develop.

The elegant mathematics of string theory faces a number of reality-challenges. The theory is marred by the consideration that at least six and possibly eleven dimensions are needed to accommodate all of the vibrational modes of the strings. (Theorists maintain that the extra dimensions have been compactified, or choked off, as the universe expanded.) Another, deeper issue is the reality-status of the strings themselves. Much like musical notes, they represent specific vibrations. But musical notes are produced by strings or by other vibrating surfaces that are part and parcel of the physical world. String theory's strings, by contrast, seem to float in geometric space-time in an unsubstantial manner reminiscent of the grin of the Cheshire cat. The body of the cat—the substance that would vibrate and by its geometrical properties determine the vibrational modes—is seldom speculated upon, yet it cannot be pure geometry: a theoretical construct cannot vibrate.

Rather than affirming that it is the geometrical fabric of space-time that undergoes pinching and tearing as the Calaby-Yau shape defines blackholes, wormholes, or elementary particles, it is more realistic to assume the presence of an underlying physical domain, a subquantum-level field, that produces the manifest effects.

An underlying field is also implied by the kind and level of coherence found in cosmology. In a realistic perspective a physical medium is indicated as the carrier of the information that correlates the diverse, and often extremely distant, parts of the universe. The large-scale uniformity of the cosmic macrostructures and, as we shall see, also the astonishingly precise fine-tuning of the universal constants, argue for the presence of a cosmically extended correlation-transmitting field.

Biological Fields

The concept of an interconnecting field conserves its relevance as we shift from the physical to the living world. The affirmation of a field as a basic element in the living organism is not new. As early as 1925, Paul Weiss, inspired by Wolfgang Koehler's Gestalt theory, applied the field concept to processes of limb regeneration in amphibians, and later he generalized the concept to all forms of ontogenesis. On the basis of his experimental work Weiss concluded that the emergence of organs and tissues during development indicates that the emerging parts assume patterned spatial relations exhibited in geometric features of position, proportion, and orientation. These, he said, are "field actions." Each species has its own "morphogenetic field," and each individual's morphogenetic field is a nested hierarchy of subsidiary fields.

Likewise in the 1920s Alexander Gurwitch noted that the role of individual cells in embryogenesis is determined neither by their own properties nor by their relations to neighboring cells, but by a factor that seems to involve the entire developmental system. This, he said, is a system-wide force field created by the mutual effect of the individual force fields associated with cells. The boundaries of the field of an embryo, for example, do not coincide with the boundaries of the embryo itself: they penetrate beyond it. Embryogenesis, Gurwitch said, occurs within the embryo's morphogenetic field.

In 1934 Conrad Waddington introduced the idea of "individuation fields" active in the formation of organs, and later extended the field-idea to "chreods," the developmental pathways of embryogenesis (Waddington 1966).

Although biological field theories were pioneered in the 1920s and attained wide popularity in midcentury, the physical properties of the fields were not well defined and in subsequent decades interest in them declined. In embryology, for example, biochemical methods did not enable researchers to discover the nature of the fields that would govern limb polarity, neural patterning, lens induction, and other developmental processes. Field concepts came to be regarded as speculative, and in the last decades of the twentieth century only a handful of investigators persisted in producing biological field theories. For the most part, biologists shifted their attention to the biochemistry of specific genetic mechanisms, a powerful approach that yielded a plethora of practical applications. They did not contest the presence of complex fields associated with cellular matter, but viewed the role and function of the associated electric, magnetic, and other fields as secondary effects without biological significance.

However, beginning in 1972, a group of physicists at the University of Marburg, Germany, investigated the emission of photons in living organisms and came to the conclusion that a coherent photon field is responsible for intra- and inter-cellular communication in the organism, as well as for the regulation of biological functions such as cell growth and differentiation. This concept was reinforced by the biophoton-research of Fritz-Albert Popp and collaborators at Germany's International Institute of Biophysics at Neuss. Light of extremely low intensity is involved in intercellular communication; low intensity laser occurs as well. Günter Rothe concluded that biological systems use electromagnetic as well as other fields of an as yet unknown kind in ensuring the collective coherence of the organism (Rothe 2002).

Since the late 1970s Brian Goodwin has been advocating a field approach to regeneration and reproduction, processes in which a whole is generated from a part (Goodwin 1979, 1982, 1989). According to Goodwin these processes cannot be viewed solely in terms of germ plasm and DNA, but must be seen as arising from the field properties of living organisms. Biological fields generate spatial orders that influence the activity of genes, and gene activity in turn influences the fields. The field is the unit of form and organization, while the molecules and cells that make up the body are the units of composition: fields structure them into the order that characterizes the organism. Life is a "sacred dance" of cells within organisms, and of organisms within their milieu, where biological fields keep the partners in step.

By the end of the century Scott F. Gilbert, John M. Opitz, and Rudolf A. Raff proposed to bring fields back into the mainstream of biology (Gilbert et al. 1996). In their view evolutionary and developmental biology needs to be brought together in a new synthesis in which morphogenetic fields mediate between the genotype and the phenotype. Gene products interact to create morphogenetic fields that act on the phenotype and influence the way it functions and develops. The role of morphogenetic fields has been widely discussed since the publication of Rupert Sheldrake's "hypothesis of formative causation." In this concept morphic fields are associated with all living organisms and are responsible for their self-maintenance as well as for their morphogenesis (Sheldrake 1981, 1988).

In the opening years of the 21st century the advance of natural healing and other forms of non-conventional medicine contributed to the revival of interest in biological fields (Benor

1990, 1993, 2002). Termed *energy* or *information* medicine, the increasingly widespread alternative forms of healing suggest that the electric and magnetic fields generated by tissues and organs, and altered by pathologies, are useful indicators of the body's condition and serve for purposes of diagnosis as well as of healing. These fields appear to be essential elements in the communication of cells within the organism, and of the organism with its environment.

Investigating the physical basis of acupuncture, Chang-Lin Zhang, of Zheijang University, China and Siegen University, Germany, found an electromagnetic field composed of interference patterns of standing waves in the resonance cavity of the human body, supported by the energy flow of the organism. The field of interference patterns appears to be holographic, in that changes in the conductivity of the measurement current (which is proportional to the field inside the body) appear simulatenously not only on all acu-points but, to a lesser extent, on every point of the skin. Electrical resistance changes occur as soon as the organism undergoes a pathological, physiological, or even psychological change. Zhang concluded that the discovery of an invisible electromagnetic field within the body offers not only a new understanding of the background of acupuncture and other forms of holistic medicine, but also a quantitative evaluation of the degree of coherence of the body-mind system (Zhang 2002).

In reviewing the state of the art in non-conventional medicine, James Oschman suggested that, in most cases, biomagnetic rather than bioelectric fields are likely to be involved in cellular and organic communication since they do not attenuate significantly when passing through body tissue (Oschman 2001).

The rediscovery of biological fields as basic elements of organic functioning makes for a fundamental shift in emphasis. It is similar to the "figure-ground switch" described by Gestalt psychologists, where the visual perception of an image is snapped back and forth between seeing one of its aspect as figure and another as ground. In mainstream biology the figure is the assembly of organic molecules constituting the cell, and the fields produced by cellular communication—insofar as they are taken into account at all—are seen a physiologically insignificant background. By contrast in cutting-edge research the figure is the field, and the molecules, cells, and organs on which it acts is the material context: the ground.

Transpersonal Fields

Last but not least, we should note the relevance of fields to the phenomena of transpersonal coherence coming to light in consciousness research.

As we have seen, recent experiments indicate that under certain circumstances, spontaneous nonsensory transfer of information can take place between different individuals. The resulting contents of consciousness are anomalous in light of classical materialist assumptions. Skeptics dismiss the phenomena, claiming that they are generated by the brain of the experiencing subject—upon stimulation certain parts of the brain (for example, the frontal lobes of the neocortex) have been known to generate images and impressions without sensory stimulus. If this were the case, the contents of transpersonal experiences would not refer to anything beyond the brain itself. However, one cannot exclude a priori that the brain, an ultrasensitive quantum system, would also interact with the world in a nonsensory mode. In that case the information generated in transpersonal experiences would have a veridical aspect.

The assumption of a flow of information between the brain and consciousness of individuals beyond the range of their sensory organs does not call for esoteric assumptions; it is sufficient to contemplate the presence of a physically real information-transmitting field accessed by the supersensitive quantum brain. (The standard objection, that consciousness does not exist in space, while "field" is a spatial construct, does not hold: the concept of field does not apply to consciousness directly, but to the brain with which consciousness is associated.) The currently advanced quantum brain theories affirm that the brain, itself a macroscopic quantum system, receives and processes signals not only from the body, and not only through the body's sensory organs, but also through its sensitivity to the quantum fields that surround the body. Roger Penrose, for example, who rejects the possibility of human consciousness being simulated by Turing machines or other artificial intelligence methods, locates the specificity of consciousness in the large-scale quantum coherence of the brain (Penrose 2000). Consciousness, he suggests, is the manifestation of an entangled cytoskeletal state and its involvement in the interplay of classical and quantum states. It involves the process of "objective reduction," a self-organizing collapse of the quantum wavefunction related to instability at the most basic level of space-time. Large areas of the brain are nonlocally coupled, so that quantum nonlocality and entanglement extend into macroscopic dimensions and involve the whole brain together with the environment around the brain.

Stuart Hameroff concluded that at the leading edge of contemporary brain research consciousness is seen as brain activity coupled to self-organizing ripples at a fundamental level of physical reality (Hameroff 1998).

PART 2

The Connectivity Hypothesis

In part 1 we presented evidence that the anomalous forms of coherence—intrinsic, instantaneous and distance-independent correlations among the proximal and distant parts of a system—are coming to light independently in areas of investigation as diverse as quantum physics, physical cosmology, the biophysics of the organism, evolutionary biology, and consciousness research. We then argued that these phenomena imply system-wide connectivity in nature, and that such connectivity suggests in turn the presence of an interconnecting field.

Fields—physical, biological, and transpersonal—have been postulated in all of the pertinent disciplines. It is possible, and in view of the overall consistency of nature also probable, that we have to deal not with specific distinct fields, but with a fundamental field that produces specific physical, biological, and transpersonal effects. One and the same fundamental field may produce analogous but locally differentiated effects at different scales of size and complexity, in systems consisting of different components.

It is this possibility that we explore here. In chapters 5–9 we first outline the premises and the postulates of the "connectivity hypothesis" that can account for the currently anomalous forms of coherence, then put forward the hypothesis itself, and finally sketch out the "integral quantum science" that will emerge when the coherence presently discovered in nature becomes a basic element of our knowledge of quantum, cosmos, life, as well as consciousness.

Chapter 5

Premises

Recent findings regarding the fundamental properties of physical nature constitute the premises of the hypothesis of connectivity. The relevant findings can be grouped under three headings:

1. Space is an energy-filled plenum.
2. The energies that fill space are the virtual energies of the *cosmic plenum* (misleadingly named quantum vacuum).
3. The universal forces and constants of nature arise in the interaction of the virtual energies of the cosmic plenum with particles and systems of particles in space and time.

The First Premise: *Space is an energy-filled plenum*

The idea of space as a plenum filled with energy has been intuitively affirmed in almost all of the classical cosmologies. In classical Greece this intuition was contested by Democritus who required truly empty space—the "Void"—so that atoms, the ultimate building blocks of physical reality, could move around. However, Aristotle pointed out that what appeared to be empty space was in fact a medium that conducted heat as well as light.

At the dawn of modern science Newton's concept of space as a passive receptacle for the motion of mass points coincided with the idea of Democritus. But in the course of the nineteenth century this concept conflicted with Maxwell's theory of electromagnetism, where electromagnetic phenomena are due to the action of a field that extends throughout space. The field carries electromagnetic waves, including visible light, much like water carries waves.

In nineteenth century physics the space-filling field responsible for electromagnetic wave-propagation was conceived as the luminiferous ether. The theory advanced by Jacques Fresnel predicted mechanical friction as bodies move through the ether. Yet in the famed Michelson-Morley experiments at the turn of the twentieth century such friction—which should have slowed the propagation of light—failed to materialize. The speed of light remained the same whether the light beam moved in the direction of the Earth's rotation or opposite to it; the movement of the planet through the ether did not influence it. The contradiction with the dominant concept of the ether was resolved in 1905, when Einstein published the special theory of relativity. In that theory the speed of light is constant, and positions relative to an absolute reference-frame, such as a stationary ether, are not required.

Within a few years the concept of the luminiferous ether was banished from the physicists' world picture. But this did not mean that henceforth space could be seen as passive and empty. A. A. Michelson himself insisted that the negative outcome of the Michelson-Morley experiments did not call into question the existence of a medium called the ether, whose vibrations produce the phenomenon of heat and light, and that is supposed to fill all space (Michelson 1881). The fact that the interpretation of the ether produced by Fresnel was disproven, Michelson noted, is not proof that there is no medium that fills space and transmits a variety of effects: gravitational, electromagnetic, and possibly still others. Subsequently Einstein himself came to this view: in a 1920 talk before the Swiss Academy of Sciences he noted that according to the general theory of relativity space is endowed with physical qualities—in this sense, he said, there is an ether.

In the last few years this insight gained fresh recognition. The long-standing neglect of a medium that fills space and time and transmits a variety of effects is being reconsidered. This does not bring back the nineteenth century concept of the luminiferous ether; instead, physicists are reviving the idea of space-time as a physically real energy-medium. This concept has been foreshadowed in concepts and theories advanced not only by Einstein, but also by Planck, Born, Jordan, Heisenberg, Casimir, Lamb, Rutherford, Dirac, Pauli, Weisskopf, and Yukawa, and more recently by Wheeler, Bohm, Heim, Yang and Mills, Higgs, Goldstone, Sinha, Sudarshan, and Vigier. Today it is reinforced by the surprising findings of physical cosmology. Space, it appears, is far from empty: it is filled with highly esoteric physical realities, partly material, partly energetic. According to the conception now coming to light, cosmic space contains per weight only 4 percent of baryons. Twenty-three percent is "dark matter," and seventy-three percent is "dark energy."

The Second Premise: *The energies that fill space are the virtual energies of the cosmic plenum*

The twentieth century evolution of physics superseded the classical definition of the vacuum: it is not just the lowest energy state of a system of which the equations obey both wave mechanics and special relativity, but a physically real energy domain. This domain is neither a vacuum, nor indeed a quantum vacuum, but a plenum that extends throughout the cosmos below the level of quanta: the *cosmic plenum*.

The energy density of the cosmic plenum is staggering. When the wavelength of electromagnetic radiation is taken to be equivalent to the diameter of the nucleon (approximately 10^{-35} m), the radiation's frequency rises to 10^{44} Hz. Then, in light of Einstein's mass-energy relation, the energy density of the vacuum/plenum turns out to be of the order of 10^{93} kg/m^3. This is greater by a factor of 80 than the energy density of the atomic nucleus (the latter is 10^{13} kg/m^3).

The cosmic plenum is a virtual plasma, consisting of energy fields fluctuating around their zero baseline value. As the energy of these fields remains active even at absolute zero temperature, the energy that fills space came to be known as zero-point energy (ZPE).

The ZPE concept originated in 1912, when Planck derived it from his theory of blackbody radiation. Although Planck himself abandoned it—he believed that zero-point energies could not have observable consequences—Einstein later found that Planck's radiation formula did require the existence of such energies. Zero-point energies are part of quantum electrodynamics (QED) through the work of Born, Heisenberg, and Jordan, but for reasons of computability zero-point fluctuations are "renormalized"—eliminated from the equations. In SED (stochastic electrodynamics), however, zero-point energy fluctuations are taken explicitly into account and are shown to be at the root of a number of quantum effects.

Zero-point fluctuations create and annihilate virtual photons. The virtuality of the photons is due to their extreme short life expectancy: the interval between their emergence from, and reabsorption in, the plenum is too small to be observed and measured. But, though the photons are virtual, unlike "real" photons they have mass, and the shorter their duration the greater their mass.

Photons are not the only virtual particles in the cosmic plenum. While virtual photons are exchanged by electrons in electromagnetic interactions, there are virtual positrons surrounding the electrons. In addition to bosons (photons, gravitons, and gluons), also fermions (protons and neutrons) fluctuate in the plenum. The encompassing "Dirac-sea" was proposed by Paul Dirac in 1930 to explain why matter is stable in transitions from positive to negative energy states. In his conception the vacuum is a physical state in which all (and only) negative energy levels are present. This does not produce observable charge density. As the subsequent generalization of his theory specified, the zero-charge of the fermion vacuum/plenum is due to the mutual cancellation of the charges of particle-antiparticle pairs.

Attempts to unify nature's principal forces entailed a progressive reconceptualization of the physical properties of the quantum vacuum. In unified and grand-unified theories the concept of this cosmic energy-sea transformed first from classical physics' ether-filled space into the electromagnetic zero-point field (ZPF). Then, in quantum field theory, the electromagnetic ZPF evolved into the fermion vacuum of the Dirac-sea; and in "super-grand unified" theories the fermion vacuum transformed in turn into the "unified vacuum"—the cosmic plenum underlying such multidimensional constructs as supersymmetry and supergravity.

The Third Premise: *The universal forces and constants of nature arise in the interaction of the virtual energies of the cosmic plenum with particles and systems of particles in space and time.*

Interactions between of the cosmic plenum with particles and systems of particles in space and time come to light in some of the most advanced, if as yet also most speculative, theories of avant-garde physics.[1] It appears that many of the fundamental interactions in the physical universe are mediated by particles embedded in the cosmic plenum. Electric interactions are mediated by virtual photons: they affect the magnetic field. Magnetic interactions are mediated by real photons, affecting the electric field, and gravitational interactions are conveyed by virtual gravitons, acting on the gravitomagnetic field. Gravitomagnetic interactions are conveyed by real gravitons that alter the gravitational field, weak nuclear interactions are mediated by W+, W−, and Z_o bosons, and strong nuclear interactions are mediated by virtual gluons.

Even the masses of particles may arise in interaction with the cosmic plenum. The so-called Standard Model of particle physics has no mechanism that would account for the masses of elementary particles. It is generally believed that the masses are generated by interaction with a universal field: the stronger the interaction of a particle with the field, the greater its mass. The exact nature of the pertinent field is not known; it is often held to be a variety of scalar field, possibly the elementary field known

---

1. Here and throughout this exposition the expression "particles" refers to the smallest identifiable units of physical nature—not point-particles, but quantized energy-packets that at the Planck-dimension can be conceptualized as vibrating strings or filaments.

as Higgs field. The interaction of particles with the fields postulated in the Standard Model together with a universal scalar field is said to be responsible for particle masses.

In order to complete the Standard Model the existence of the pertinent scalar field—or possibly of various scalar fields—needs to be confirmed. As Stephen Weinberg noted, this may be achieved in 2020, when the Large Hadron Collider at CERN, the European nuclear research facility, comes on line.

The assumption in regard to the interaction of the cosmic plenum with elementary particles derives from relativistic quantum field theory. Here both boson and fermion fields are postulated, and the vacuum is considered the ground state (the state of minimum energy and maximum stability) of the many-field system. In the ground state, the vacuum's boson and fermion fields are "virtual." When, however, a region of the vacuum is excited—by the influx of energy of the order of 10^{27} erg/cm^3—a pair of particles is created, of which the "real" twin can establish itself in spacetime. In its place a "hole" of positive charge and mass remains (the Dirac-sea has negative charge and negative mass).

Dirac showed that fluctuations in fermion fields produce a polarization of the vacuum, whereby it affects the particles' mass, charge, spin, or angular momentum. However, the effect on particles is extremely subtle; its full extent is only now coming to light.

The systematic investigation of vacuum effects began in 1948, when H. G. B. Casimir discovered the effect named after him, together with the force responsible for it. Between two closely placed metal plates some wavelengths of the plenum's energies are excluded, and this reduces its energy density with respect to the energies on the outer side of the plates. The disequilibrium creates a pressure that pushes the plates inward and together. In 1997 S. K. Lamoreaux measured the force responsible for the Casimir effect to a high degree of precision. This led to major research projects that developed experimental protocols for extracting energy from the cosmic plenum—projects generally known as "engineering the vacuum."

The Lamb-shift, another already classic vacuum effect, consists of the frequency-shift exhibited by the photons that are emitted as electrons orbiting the nucleus leap from one energy state to another. More fundamental effects are also ascribed to interaction with the vacuum. Already in 1968 Andrei Sakharov, seeking to derive Einstein's general relativity formulas from a more basic set of assumptions, proposed that relativistic phenomena may be effects induced in the vacuum by the presence of matter (Sakharov 1968). In his theory the force of gravitation is more like the vacuum-generated Casimir force than the Coulomb force between charged particles, the force with which it is usually compared. A few years later Paul Davies and William Unruh put forward a hypothesis that differentiates between uniform and accelerated motion in the vacuum's zero-point field. Uniform motion would exhibit the ZPF as isotropic, whereas accelerated motion would produce a thermal radiation that breaks open the directional symmetry.

Sakharov's pathbreaking theory, together with the Davies-Unruh hypothesis, gave rise to a series of research efforts that constitute a rapidly growing sector of fundamental physics. Here we review a sample of avant-garde research into the fundamental field that, though still termed "quantum vacuum," constitutes a cosmic plenum that interacts with particles in space and time.

A Sampling of Recent Theories of Vacuum (Plenum) Interaction with Particles in Space and Time

László Gazdag of the Science University of Pécs in Hungary followed up his internationally known countryman Lajos Jánossy's vacuum-interaction interpretation of relativity theory by elaborating the concept of the superfluid vacuum[2] (Gazdag 1989, 1998). The concept rests on the analogy between liquid helium and an ideal Bose gas. This accounts for the puzzling fact that light propagates in space as a transverse wave. Transverse waves, it is known, can only propagate in solid matter, such as crystals. At the same time celestial bodies move through space without apparent friction. This means that the plenum acts both as a supersolid crystal and as a superfine gas. The solution to this

---

2. Gazdag's reformulation of the equations of General Relativity is in Appendix I.

apparent paradox is to view the plenum as the dynamic analogue of superfluid helium. Near absolute zero on the Kelvin scale helium gas is both supersolid and superfine; it consists of bosons rather than electrons. The theory of this phenomenon was given by J. Bardeen, L. N. Cooper, and G. R. Schrieffer in 1957: in a superconducting medium electrons constitute duplets. Two half-spin electrons (which are fermions) behave as one full-spin boson. Accordingly, the vacuum can be considered a superdense boson-gas.

Two-fluid hydrodynamics is the indicated theoretical model. The energy structure of the superfluid vacuum is differentiated into a "collective coherent" and a "fluctuating coherent" ground state, separated by an energy gap. The latter integrates stability with dynamism and protects the collective ground state against perturbations. As Charles Enz showed in 1974, this model has wide significance: it describes order in condensed matter in general (dielectric crystals, magnetic crystals and plasmas), as well as the behavior of superconductors and superfluids.

In a superfluid vacuum the structure of space-time is determined not merely by the presence of mass, as in Einstein's general relativity, but by real and virtual bosons underlying the known varieties of physical interactions. When the bosons flow uniformly, the structure of space-time is flat—it is Euclidean. When the plenum's boson fields move in a nonuniform manner (deviating from a straight line, accelerating, or decelerating), the structure of space-time is deformed: it becomes Riemannian. Then the plenum—and hence space-time—loses its superfluid properties and various interaction effects become manifest. The gluon field's nonuniform motion creates the strong and weak interaction forces within the atomic nucleus. If the graviton field moves nonuniformly, the gravitational force appears; if the electromagnetic bosons flow in that fashion electromagnetic phenomena emerge. And when the electromagnetic bosons flow at the velocity of the light constant c, a light beam (a stream of photons) is produced.

While subrelativistic uniform motion in the vacuum is frictionless—it does not produce any effect on moving objects—at speeds approaching the velocity of light Einstein's relativistic effects surface: the slowing down of clocks, the increase of weights, and the contraction of measuring rods. These are seen as physical effects, due the interaction of objects with the vacuum.

Other innovative theories of the physical vacuum are the work of Manfred Requardt of the University of Göttingen in Germany and, independently, of Ignazio Licata of the University of Palermo in Italy (Licata 1989, Requardt 1992). In their conception quanta are coarse-grained manifestations of "physical space-time." In Licata's theory physical space-time is reticular, functioning as the ultrareferential structure in which absolute deformations are described by the stochastic metric tensor and express deviations from isotropy and homogeneity in the Lorentz-invariant background. Lorentz transformations are real effects created by the motion of matter in the subquantum domain of reticular space-time.

A theory put forward by Bernhard Haisch, Alfonso Rueda, and Harold Puthoff claims that the force of inertia itself is a product of the interaction of charged particles with the ZPF of the vacuum (Haisch et al. 1994, Puthoff 1987, 1989a, 1989b, 1993). In this conception inertia is a vacuum-based Lorentz-force operating at the subparticle level and creating opposition to the acceleration of material objects. This "electromagnetic drag force" is acceleration-dependent due to the spectral characteristics of the zero-point field. The inertial mass m_i is derived starting from the consideration that in stationary as well as in uniform-motion frames the interaction of a particle with the ZPF results in random oscillatory motion. Fluctuating charged particles produce a dipole scattering of the ZPF, parametrized by the scattering spectral coefficient $h(w)$, which is frequency-dependent. Because of the relativistic transformations of the ZPF, in accelerated frames the interaction between a particle and the field acquires a direction: the scattering of ZPF radiation generates a directional resistance force. This force is proportional to, and directed against, the acceleration vector for the subrelativistic case. It turns out to have the proper relativistic generalization. It thus appears that inertia is not a property of quantized particles, but a product of their interaction with the zero-point field of the vacuum.

The theory that inertia is a product of vacuum interaction has further entailments. If the force of inertia originates in the interaction between the zero-point field and charged particles, then the principle of equivalence between this force and the force of gravitation requires that gravity should so originate as well. In that event gravity is a force generated in ZPF-charge interactions, much as Sakharov foresaw.

Haisch, Puthoff, et al. maintain that the vacuum-interaction interpretation of gravity is analytically equivalent to the general relativistic treatment of gravity as space-time curvature. It is based on a different process, however. In the new concept the electric component of the zero-point field causes charged particles to oscillate; the oscillation gives rise to secondary electromagnetic fields. A given particle experiences both the electric forces of the ZPF, which cause it to oscillate, and the secondary forces that are triggered in the field by another particle. The secondary forces generated by the second particle act back on the first particle. The net effect is an attractive force among the two particles. Gravity is thus a long-range interaction force among particles, much as the van der Waals force. Inertia, gravity, as well as the van der Waals force are field effects generated by the excitation of the vacuum.

The interpretation offered by Haisch *et al.* consists of two parts. In the first part the energy of the ultrarelativistic oscillations known as *Zitterbewegungen* is equated to gravitational mass m_g, after dividing by c^2. Except for a factor of 2, this produces a relationship between the gravitational mass and electrodynamic parameters identical to the postulated inertial mass m_i. It can be shown that the gravitational mass m_g should be reduced by a factor of 2, which would then yield a strict equivalence between m_i and m_g, that is, between the forces of gravitation and inertia. The second part of the analysis derives an inverse square force of attraction from the van der Waals force-like interaction between two driven oscillating dipoles. This analysis is admittedly incomplete: it requires further development in the framework of a fully relativistic model.

The thesis that inertia and gravitation are vacuum interaction products, though speculative, is supported by Gazdag's revised relativity theory. Gazdag traces inertia and gravity to the interaction between massive particles and the superfluid vacuum. The dynamics of action in the two cases are analogous. In regard to inertia, the field, which loses its superfluid properties through the accelerated motion of massive particles, resists their acceleration; while in the case of gravity it is the accelerating field that acts on massive particles. When superfluidity ceases, macroeffects are manifested, and the inertial, the gravitational, and the electromagnetic masses appear.

The hypothesis of gravity as a product of interaction with the vacuum—more precisely, as an attractive force generated in the vacuum by massive particles—coincides with independent experimental findings by Dezső Sarkadi and László Bodonyi (Sarkadi and Bodonyi, 1999). These investigators replaced the usual Cavendish-type torsion balance for the measurement of the gravity-constant with a large and heavy physical pendulum of vertical bell shape and stiff frame. In experiments with this pendulum the source of the compensation momentum is the stable gravitational field of the Earth, which gives a more accurate and reliable measurement than the torsion method. In numerous experiments Sarkadi and Bodonyi showed that, while in the case of unequal masses the gravitational force satisfies the usual estimates, when the mass of the objects approaches the same value, gravitational attraction between them diminishes. Among objects of commensurable mass, gravity is reduced, and an extrapolation of the rate of reduction suggests that in the case of objects of precisely equivalent mass, gravity vanishes. The investigators suggest that Newton's law of gravitation may be valid only for point-like masses; in the case of extensive masses it leads to improper results.

Gravity, it appears, is neither an intrinsic property of massive objects (or of "matter," as Newton and Ernst Mach speculated), nor the result of space-time curvature (as in Einstein's general relativity). Rather, it is generated between extended masses in the vacuum. When the masses are highly unequal, the vacuum field is deformed and creates attraction between the objects. When the masses are equal, the field is not deformed and the force does not appear. This finding has revolutionary consequences, for it questions the validity of the gravity-postulate of general relativity: gravity in this light is not a property of the geometry of space-time, but the consequence of interaction between extended masses and the vacuum field.

In addition to gravity, inertia, and mass, the Planck constant has also been attributed to interaction between massive particles and the vacuum. Here interaction is assumed between the virtual bosons of the vacuum and material oscillators (fermions). It is known that at higher levels of oscillation only multiples of a basic quantity are given, that is, the effect constant is quantized. (As an effect- rather than energy-quantum, the Planck constant [of which the value is $h = 6.625 \times 10^{-34}$ Js] is the differential between kinetic and potential energy, the integral of the Lagrange function. It is the lowest possible kinetic and potential energy time-integral in the oscillation of an oscillator.) According to Gazdag, the effect constant results from the interaction of the oscillator with the vacuum. In any finite period of time an oscillator interacts either with one vacuum boson, or with its integral multiple. Thus the reason why the Planck effect-constant is quantized is that energy transport takes place within a field structured by quantized bosons.

The theory put forward by Judah Tzoref is based on an analogous insight. It interprets matter-free space as having an inherent vacuum intensity (Tzoref 1998, 2001). In Tzoref's "vacuum kinematics" fundamental particles and interactions are manifestations of the geometry and kinematics of the vacuum. In vacuum kinematics all particles, energy, and interactions are described as manifestations of spatial and temporal changes in vacuum geometry. "Vacuum forms" (particles) constitute "vacuum states" (fields), which induce a "vacuum response" (interaction). The mathematics based on this concept allow an interpretation of motion, electrodynamics, gravity, as well as the strong and the weak nuclear forces as vacuum responses to irregularities induced by particles (as vacuum forms) enfolded in fields (as vacuum states).

Interaction between particles and the vacuum through secondary fields generated by charged particles is the basis of yet another theory, by the Russian physicists G. I. Shipov, A. E. Akimov and colleagues. They suggest that vortices are created in what they term the "physical vacuum" by the spin of charged particles (Akimov et al. 1997, Akimov & Tarasenko 1992, Shipov 1998). The theory modifies and generalizes Dirac's electron-positron model: the vacuum's energy field is considered a system of rotating wave packets of electrons and positrons, rather than a sea of electron-positron pairs. Where the wave packets are mutually embedded, the torsion field is electrically neutral. If the spins of the embedded packets have the opposite sign, the system is compensated not only in charge, but also in classical spin and magnetic moment. Such a system is said to be a "phyton." Dense ensembles of phytons approximate a simplified model of the torsion field in the physical vacuum. Its vortices are information carriers, linking physical events at a group-speed of the order of 10^9 c, that is, one billion times the speed of light.

When the phytons are spin-compensated, their orientation within the ensemble is arbitrary. But when a charge q is the source of disturbance, the action produces a charge polarization of the vacuum, as prescribed by quantum electrodynamics. When a mass m is the source of disturbance, the phytons produce symmetrical oscillations along the axis given by the direction of the disturbance. The vacuum then enters a state characterized by the oscillation of the phytons along their longitudinal spin-polarization; this is interpreted as a gravitational field (*G*-field). Given that the gravitational field is characterized by longitudinal waves, it cannot be screened, which is in accordance with observation and experiment. Thus the gravitational field is said to be the result of vacuum decompensation arising at its point of polarization, a notion that corresponds to Sakharov's theory. In this conception *m*-disturbance produces the *G*-field, much as *q*-disturbance produces the electromagnetic field.

The Russian physicists represent the vacuum equations in the spinor form and obtain a system of nonlinear spinor equations where two-component spinors represent the potentials of torsion fields. These equations, as Roger Penrose has shown, describe charged and neutral quanta as well as classical particles. They allow the vacuum field to be disturbed not only by charge and mass, but also by classical spin. In that event the phytons oriented in the same direction as the spin of the disturbance keep their orientation, while those opposite to the spin of the source undergo inversion. Then the local region of the vacuum transits into a state of transverse spin polarization. This gives the "spin field" (*S*-field), a condensate of fermion pairs. Accordingly the vacuum is a physical medium that assumes various states of polarization. In charge-polarization it is manifested as the electromagnetic field; in matter-polarization it gives rise to the gravitational field; and in spin-polarization it constitutes the nonlocal spin-field.

A number of speculative theories consider the vacuum as the unified ground of matter and life. There is, for example, Alex Kaivarainen's Unified Model (UM), where the concept of "Bivacuum," composed of nonmixing subquantum particles of opposite energies, constitutes the unified dynamic matrix of the universe (Kaivarainen 2002). The UM attempts to unify vacuum, matter and fields from a few basic postulates. In the Bivacuum nonmixing subquantum particles of opposite energy form vortical structures. Named Bivacuum fermions and antifermions, these structures have an infinitive number of double cell-dipoles, with each cell containing a pair of correlated rotors and antirotors of opposite quantized energy, virtual mass, charge, and magnetic moment. In O. E. Wagner's "W-wave" theory, in turn, longitudinal W-waves (so named because they were first found in wood) radiated by various forms of living and nonliving matter, create a response from the vacuum in the form of an automatic return wave. Much of the organization of the universe and of life, Wagner maintains, is due to the standing waves created by the response of the vacuum to longitudinal W-waves (Wagner 1999a, 1999b).

Michael J. Coyle maintains in turn that the quantum vacuum is a kind of neo-ether through which all matter and energy evolves in continuous nonhertzian communication. The neo-ether model posits that the vacuum state has the same properties as a dynamic holography crystal: it infinitely stores the diffraction patterns of matter-energy systems in a nontemporal and nonlocal fashion analogous to holographic information storage throughout the recording material. Due to quasi-infinite superposition possibilities, the quantum vacuum is the source of all possible matter-energy states and parallel universes. Individual states can be selected using a fundamental form of a holographic, self-referencing, phase-conjugate mirror-like process. This process accounts for the nonlocality observed in biological organisms as well as in consciousness. Thus matter coupled with time is a discontinuous teleportation of information across a sub-Planckian void. All possible superimpositions of particles or parallel universes can be selected at the point of information-transfer, since they all share information through interference at that point, prior to becoming actualized in a given universe (Coyle 2002).

Although still speculative, such exploratory work indicates that cutting-edge particle and field physicists view the energies that fill space as a physically real, cosmically extended plenum. This cosmic plenum interacts with the particles and systems of particles that make up the observable universe. It appears that it is the interaction between particles and the cosmic plenum that determines the mass of the particles, as well as the values of the universal forces and constants of nature.

Chapter 6

◆

Postulates

The premises of connectivity hypothesis are based on the current finding that cosmic space constitutes a physically real plenum that, through interaction with particles and systems of particles, determines many and conceivably all of the physical properties of nature.

The postulates of connectivity hypothesis build on this finding and make a further claim. They suggest that interaction between the energies that fill cosmic space and charged particles and systems of particles not only determines the mass of the particles and the values of the fundamental forces and constants, but also correlates the particles and systems of particles, creating coherence among them.

The First Postulate: *The zero-point field of the cosmic plenum is a complex field that includes a field of scalar potentials.*

The first postulate suggests that the zero-point field is not the classical electromagnetic field, but a complex field of effective magnetic vector potentials and equally effective electrostatic scalar potentials.

Scalar potentials were part of the original theory of electromagnetism: Maxwell's theory was based on the Quarternion-algebra developed by W. R. Hamilton, where the electromagnetic field has both a magnetic vector-potential and an electrostatic scalar-potential (Maxwell 1873). However, in the subsequent revision of that theory by Oliver Heaviside, Josiah Gibbs, and Heinrich Hertz electric and magnetic potentials were not viewed as physically real components of the field, and the scalar terms in Maxwell's equation were omitted. This was well suited to a variety of technical applications, but it ignored the subtle yet physically real effects produced by longitudinally propagating waves such as scalars. The longitudinal component of the vacuum was predicted already by Schrödinger, but fell into disfavor due to its conflict with some aspects of quantum theory, such as gauge invariance. There is increasing experimental evidence, however, that the longitudinal component is real: transverse electromagnetic waves are accompanied by longitudinal scalar waves.

Evidence for the existence of nonhertzian scalar waves surfaced with the analysis of the Aharonov-Bohm effect, where E and B, the field's magnetic and electric components, are zero (so that the field does not convey actual force), yet the potentials are nonzero (Aharonov & Bohm 1959). In this experiment an electron is confined in a Faraday cage so that it experiences the potential of the electromagnetic field to which it is exposed, but not the field itself. Under irradiation the electron undergoes a phase-shift even though it is not experiencing the EM field. The shift can be detected when a coherent electron beam is split into two and each split beam shifts its phase differently. On recombining, an interference pattern results. T. T. Wu, C. N. Yang, and T. W. Barrett have shown that in these cases the $U(1)$ symmetry of the electromagnetic field is "conditioned into" the higher $SU(2)$ form. The potentials control the phase of the electromagnetic fields and thus mediate between these fields and matter-energy systems on the one hand, and the quantum vacuum (plenum) on the other (Wu & Yang 1975).

Already at the end of the 19th century Nicola Tesla noted that nonhertzian scalar waves are fundamental in nature. He claimed to have proven the existence of chargeless and massless particles that propagate longitudinally and whose velocity is not subject to the light-constant c. Although Tesla's theories have long been disregarded, the scalar waves named after him can be, and frequently are, generated with Caduceus, Moebius, and similar coils. This allows electromagnetic waves of contrary vector potential to cancel each other. As the vector potential goes to zero, the field's energy vanishes. Nevertheless, the canceled waves continue to carry information; they affect phase rather than amplitude. It appears that an inner structure and a corresponding dynamic are imposed on the null-vector field. Two null-vector fields may carry entirely different information and produce correspondingly different effects. A number of investigators, including William A. Tiller, Thomas Bearden, V. K. Ignatovich, and R. W. Ziolkowski, revise Heaviside's standard electromagnetic theory and bring scalars back into physics. Scalars are coming back also into biophysics: recent experiments indicate that vector and scalar potentials affect living tissue, producing effects that are distinct from those produced by electromagnetic fields (Rein 1989, 1997, 1998, Ho et al. 1994, Smith 1994).

The first postulate of connectivity hypothesis agrees that the scalar component of the cosmic plenum produces a physical effect on the motion of charged particles and systems of particles. It considers charged particles soliton-like wave-packets arising through the excitation of the plenum. Excitation produces perturbation: classical electromagnetics predicts that a fluctuating electric charge emits an electromagnetic radiation field. The energy associated with the electromagnetic field gives rise to waveform fluctuations—*Zitterbewegungen*—that propagate at or near the speed of light. In standard electromagnetic theory only vector-potentials are taken into account, which generate the transverse electromagnetic waves; however, the radiation field created by the excitation of the plenum also has a physically effective longitudinally propagating scalar component. This component affects phase, rather than amplitude. Longitudinal excitations propagate through the plenum similarly to waves propagating in a body of water. Much as water fills the sea, the scalar field pervades space, and excitations propagate through that field at variable finite velocities.

According to Bearden and other investigators, the scalar component of the plenum is more fundamental than its electromagnetic component: the latter may consist of scalar elements.

The Second Postulate: *Information in the scalar component of the cosmic plenum is in the form of interference patterns that code the collective wavefunction (Ψ function) of charged particles and systems of particles.*

The second postulate claims that excitations in the scalar field of the plenum code and convey the collective wavefunction (Schrödinger Ψ function) of the charged particles whose presence excites its ground state. The wavefunctions are sequentially integrated, with the wavefunctions of higher-order ensembles integrating the wavefunctions of lower-order ensembles.

To elucidate this proposition we should note that the wavefunction is the waveform equivalent of the quantum state of particles. As Jean Baptiste Fourier has shown, any quantitatively definable three-dimensional state can be translated into its waveform equivalent. Thus the second postulate means that the Fourier transform of the quantum state of charged particles is coded in the form of interference patterns in the field constituted of the scalar component of the plenum.

In order to grasp this porposition a further specification is in order. The quantum state defined by the Ψ function is not the classical, determinate state. Rather, this function indicates all the possible states a given particle can occupy in space and time; the actual state of the particle selects one from among these intrinsically equiprobable states. Consequently the wavefunction describes a deeper domain of reality: the domain of potentiality. It refers to the unobservable, nonmanifest world of instantaneous correlation. With the collapse of this function—as a consequence of an act of measurement or another form of interaction—events in this underlying domain emerge into the domain of actuality.

In this interpretation the deeper domain of potentiality is constituted as a scalar field in the cosmic plenum. Interaction with this field completes the physical description of the state of charged particles and systems built of such particles. Hence this writer named the scalar field of the plenum "Ψ field" (Laszlo 1993, 1994, 1996).

Current experiments indicate that in superfluid helium, which may be the closest analog of the Ψ field, minute structures arise in the form of quantized vortices. These structures propagate without friction and can carry information, either through their structure, or through their magnetic momentum. According to the second postulate analogous structures arise in the scalar field of the plenum, and these structures carry information on the charged particles and systems of particles that excite the plenum's ground state. This information is the wavefunction—the wave-transform of the quantum state—of the pertinent particles and particle systems.

The Third Postulate: *The interference patterns that code the wavefunction of charged particles extend throughout the range of scalar wave propagations in the cosmic plenum and endure indefinitely in time.*

As noted in part 1, the correlations underlying the anomalous forms of coherence in nature extend quasi-instantly across space and endure without known limitation in time. This is the foundation of the third postulate of connectivity hypothesis. It draws on the finding that, unlike classical electromagnetic waves, the propagation of scalar waves is neither limited by the light constant c, nor is it subject to attenuation proportionately to distance.

In a seminal paper published in the beginning of the twentieth century, E. T. Whittaker showed that the propagation velocity of longitudinal waves such as scalars is proportional to the mass-density of the medium in which they propagate (Whittaker 1903). This is also the case of sound waves: these are likewise longitudinal waves and they travel faster in a dense medium such as water than in a thin medium such as air. The velocity of scalar waves—more exactly, of the waveform perturbations that travel through the scalar standing waves that fill cosmic space—is proportional to the square root of their specific frequency, where the pertinent parameter is the vacuum's mass-density: this defines the local electrostatic scalar potential. The potential is higher in regions of dense mass, in or near stars and planets, and lower in deep space, a variation due to the increase in vacuum flux intensity by the accumulation of charged masses. Consequently the propagation velocity of perturbations in the scalar Ψ field is not limited by the light constant c; the latter applies only to photon propagation in the electromagnetic field.

Perturbations in the scalar field of the cosmic plenum propagate longitudinally, allowing the linear wavefronts to superpose rather than penetrate through each other. This creates interference patterns that preserve phase-information. As in ordinary holograms, the information is in a distributed form, given at all points within the range of the interfering wavefronts.

Supraluminally propagating interference patterns in the scalar field of the plenum satisfy the observed space-invariance (quasi-instantaneity) of anomalous coherence over finite distances. The time-invariance (the permanence or quasi-permanence) of the coherence is satisfied in turn by the endurance of perturbations in the field. The perturbations propagate without friction, and thus subsist indefinitely—there is no physical factor in the cosmic plenum that would attenuate or cancel them. The third postulate of connectivity hypothesis states that within the propagation range of the wavefronts, the interference patterns that code the wavefunction of charged particles and systems of particles extend throughout the pertinent regions of the plenum and endure indefinitely in time.

The coded regions of the cosmic plenum expand with the propagation of the perturbations. Ultimately the regions overlap and the wavefunctions carried by them superpose, creating higher-order wavefunctions. When all coded regions have met and their wavefunctions are superposed, the Ψ field carries the wavefunction of the universe.

A medium that carries the wavefunction of the universe requires a storage capacity of almost inconceivable magnitude. However, it is not likely to exceed the capacity of the scalar field of the cosmic plenum. Experience with conventional holography shows that holograph wave-interference patterns can be densely superposed; according to some estimates the entire content of the Library of Congress can be encoded in a multilayered holograph medium no larger than a cube of sugar. In such a superhologram every component remains conserved and individually retrievable—in the case of a library, every letter in every volume can be read out. Given the known density of holograph information storage, and the cosmic extension of the plenum, the storage capacity of the scalar field is likely to be sufficient to encode the wavefunction of all particles from one end of the universe to the other. And, given the frictionless propagation of the perturbations that create the wavefunctions, the information coded by the wavefunctions can be assumed to endure indefinitely, from the beginning of time at the birth of this universe (or of the first universe, or cycle of universes in a metaverse) to the end of time at the degeneration of the last supergalactic black holes of this universe—or of the hypothetical last universe, or last universe-cycle, of the metaverse.

Chapter 7

♦

The Hypothesis

We can now advance the hypothesis that can ground the transdisciplinary unification of a fully developed integral quantum science: the hypothesis of (quasi-universal) connectivity. Its postulates, advanced in the foregoing chapter, suggest that the coherence-inducing response of the plenum to the presence of charged particles is scale-invariant and universal: it recurs at all levels of magnitude and affects all charged particles and systems of particles in space and time. This is consistent with the already cited findings in physics, cosmology, biology, and consciousness research. The anomalous forms of coherence obtain not only as nonlocality in the domain of quanta, but also obtain in the living world and in the universe at large.

If phenomena are analogous, the processes responsible for them may have an invariant component. We hypothesize that anomalous coherence-phenomena in the various fields of observation are specific transformations of the plenum's response to the presence of charged particles. This hypothesis respects "Occam's Razor": it does not postulate entities (in this case, fundamental fields and interactions) beyond the bounds of necessity.

Connectivity hypothesis can be stated in the general form as follows:

$$S_{n(\alpha \ldots \Omega-1)} \rightarrow \Psi\text{field}\,[\Psi(S)] \quad (1)$$

where $\quad \Psi(S) = [\Psi(S_\alpha) \ldots \Psi(S_{\Omega-1})] \quad (2)$

Here S_n stands for a system constituted of charged particles of an undefined level of organization; the Greek letters alpha and omega indicate the range of organizational levels (the level of single particles is α, and the highest level of the universe [or meta-universe] is Ω), Ψ stands for the Schrödinger wavefunction, and Ψfield denotes the scalar field of the cosmic plenum. (However, for purposes of the hypothesis Ψfield can be any field or medium of universal connectivity.)

Spelled out, these notations read, "Systems of charged particles on all levels, from the level of the single particle to the level of the universe-less-one, encode the Ψ field with their wavefunction, and the thus resulting integral wavefunction of the universe in-forms systems on all levels from the single particle to the universe-less-one."

In this context *in-formation*—a term introduced by David Bohm—involves not conventional (vectorial) energy, but the more subtle influence of scalars that affects phase only. This creates information of a particular variety: one that is physically effective without conveying manifest energy: it "in-forms" the systems that receive it.

The in-formation of systems and particles is the result of a two-way process. In one direction charged particles and systems of particles structure the plenum in their region, creating interference-patterns that code their wavefunction. In the other direction, the structured plenum—the Ψ field—in-forms the particles and systems of particles that structure it.

$$S_n^{structuring} \rightarrow \text{in-forming } \Psi\text{field} \quad (3)$$

The two-way interaction between systems and the plenum is not a zero-sum exchange of the same information cycling back and forth, for the wavefunction encoded in the Ψ field is that of

The Hypothesis

the *collective* state of the entities that created it. It is never the single particle or the single property of a particle on which the wavefunction carries information, but the collective state of the system in which the particles participate; the wavefunction corresponds to the superposition of the state of all the particles within a coordinate system. Thus when the plenum response carries the wavefunction of a set of particles, it in-forms those particles with the collective state of the higher-level coordinate system in which they participate:

$$S_n \to\!\!\lrcorner\ \Psi(S_{n+1}) \qquad (4)$$

Given that the level of the wavefunction fed back through the plenum is sequentially higher than the level of the systems the wavefunction in-forms, particles in nature are in-formed with the wavefunction of the highest-level system of which they are a part. The array $S_{(\alpha\ ...\ \Omega-1)}$ is sequentially integrated

$$S_n \to \Psi(S_{n+1});\ S_{n+1} \to \Psi(S_{n+2});\ (\text{etc.}) \qquad (5)$$

with the highest level wavefunction integrating all subsidiary wavefunctions. The integrated highest-level wavefunction sequentially in-forms the lower-level (sub)systems:

$$\begin{aligned} S_{\Omega-1} &\lrcorner\ \Psi(S_\Omega)\ ... \\ S_{n+1} &\lrcorner\ \Psi(S_{n+2})\ ... \\ S_\alpha &\lrcorner\ \Psi(S_{\alpha+1}) \end{aligned} \qquad (6)$$

The in-formation of particles with their immediate as well as mediate collective wavefunction limits their intrinsic degrees of freedom: it correlates their states. This limitation is not classically deterministic: it merely defines the set of possible states the particles, and the systems built as integrated sets of particles, can occupy within the next higher-order systems.

The specific degree of correlation corresponds to the immediacy of the in-forming wavefunction. Systems on each level are immediately in-formed with the wavefunction of the ensemble of which they are a part, and this results in entanglement: a highly manifest form of correlation. In this context a particle and a system of particles form part of the same ensemble when they have either originated in the same ensemble, or had at one time entered the same ensemble and share the same quantum state. On higher-levels "being in the same ensemble" requires being part of an organism, a community, or an ecology. On still higher levels being a part of the same ensemble means sharing the same biosphere, and ultimately the same solar system, galaxy, and metagalaxy.

We can thus conclude that particles are immediately correlated by the wavefunction of the system of coordinates in which they had originated, or that they had at one time entered. In the context of these ensembles particles exhibit the highest degree of correlation: entanglement. Particles are merely mediately in-formed with the wavefunction of the higher-level ensembles in which they may participate, such as molecules, cells, and organisms. Consequently particles within entire macroscopic systems are less manifestly correlated within these systems. In the same way, cells in an organism are immediately in-formed with the wavefunction of that organism and hence exhibit instant, multidimensional coherence; and they are mediately in-formed with the wavefunction of the ecology in which the given organism participates, with the result that the cells of diverse organisms are not immediately correlated within entire ecosystems.

The same concept of mediate and immediate relationships and the corresponding degrees of correlation applies to organisms and their social and ecological systems. The relationship of organisms to the social and ecological systems in which they participate is relatively direct, and this gives rise to a more observable form of correlation among them. (In the human world this is expressed in some degree and form of social cohesion, solidarity, and mutual identification.) By comparison the relationship of particular organisms to local and global ecosystems is indirect, and the correlation that surfaces in that broader context is less manifest. (Human ties to the environment and to the whole biosphere are generally more subtle than ties to one's fellow members in a culture or community.)

The hypothesis is completed by specifying the degrees of correlation:

$$[S_n \sqcup \Psi(S_{n+1})] > [S_n \sqcup \Psi(S_{n+2})] \gg \ldots [S_n \sqcup \Psi(S)] \quad (7)$$

The Hypothesis

The wavefunction of a first-order ensemble—the coordinate system of particles—produces entanglement, the highest degree of correlation. In-formation by second-, third-, and higher-order ensembles creates correspondingly less pronounced forms of correlation. In-formation with the wavefunction of the highest-level ensemble—which is the universe as a whole—produces the weakest degree of correlation; a correlation which, while individually not measurable, at cosmic dimensions has statistically significant effects.

The different degrees of correlation can be accounted for by the different degree to which higher-level wavefunctions factorize to their components. Quantum physics tells us that correlation between two points in a coherent field decomposes into separate self-correlations at individual regions of the field. Decomposability through the factorization of the wavefunction allows a coherent system to act as a coordinated ensemble of diverse parts, rather than a monolithic totality. Quantum physics also tells us that the collective wavefunction of an ensemble of particles does not always factorize to the individual particles in that ensemble—for if it did, the particles would never be fully entangled.

The hypothesis of connectivity generalizes this principle to all scales and levels in nature. When the wavefunction of an ensemble of particles does not factorize to the individual particles that make up that ensemble, the collective wavefunction fully correlates the state of those particles. However, the wavefunction of higher-order ensembles does factorize to some extent to its component wavefunctions. This allows a relative degree of autonomy to lower-level systems (and to individual particles in the systems) while ensuring their overall correlation. The correlation limits the degrees of freedom of the particles and systems of particles within their ensembles and creates coherence-producing correlation within those ensembles.

Chapter 8

♦

Coherence Explained

Testing the Power of the Hypothesis

Coherence suggests correlation, we said in chapter 4, and implies the presence of an interconnecting field. The postulates of connectivity presented in chapter 6 claim that this is the Ψ field, a field of scalar potentials associated with the cosmic plenum. The hypothesis of connectivity put forward in chapter 7 describes the process whereby the Ψ field codes and conveys the wavefunction of the state of particles and systems of particles, correlating them within their ensembles and thereby creating coherence among them. Connectivity hypothesis states that the physical basis of anomalous coherence in the diverse domains of its observation is the response of the cosmic plenum to the presence of charged particles and systems of particles.

Connectivity hypothesis is sound, and a valid foundation of integral quantum science if it provides the simplest consistent explanation of the coherence coming to light in the various domains of experience. To ascertain whether or not this is the case calls for testing the heuristic power of the hypothesis in the areas of observation and experiment where the anomalous forms of coherence obtain—a task that calls for sustained research in a considerable number of scientific fields. A preliminary conceptual testing can, however, be undertaken here by confronting the phenomena marshaled in part 1 with the hypothesis advanced in part 2.

We carry out this initial testing by specifying the general hypothesis of connectivity for specific application to the principal domains of anomalous coherence. These domains are the microdomain of the quantum, the macrodomain of the cosmos, and the mesodomains of life and mind. We therefore analytically segment the general hypothesis into the special "microworld," "macroworld," and "mesoworld" hypotheses.

The Microworld Hypothesis

The microworld hypothesis is represented in the following notation:

$$S_{part} \to \lrcorner \; \Psi \mathit{field} \; \Psi(S_{coord}) \qquad (8)$$

Notation (8) indicates the structuring of the plenum-based Ψ field by charged particles S_{part}, and the in-formation of the particles with the Ψ field transmitted wavefunction of their system of coordinates $\Psi (S_{coord})$. The latter entangles the particles within their system of coordinates and completes the quantum mechanics of particle interaction.

In this context we should recall that the wavefunction collapses simultaneously throughout a quantum-mechanical system: a measurement in one region of an extended system has a demonstrable effect on the state of the system in any proximal or distant region. The wavefunction's collapse is not capable of transmitting significant information if the collapse itself is random: whatever signal the collapse transmits must likewise be random. However, phenomena of anomalous coherence on the level of complex systems, and already on that of particles originating in, or sharing, the same quantum state, indicate that significant information is transmitted between the various parts of the system, and this suggests that the collapse of the wavefunction of particles that share the same coordinate system is not entirely random: it is correlated within that system. Such correlation argues for the presence of a field as the physical medium of correlation-generating connectivity. Connectivity hypothesis claims that the pertinent field encodes and transmits information that corresponds to—being the Fourier transform of—the wavefunction of the given particles: that is, it conserves and conveys their angular momentum, charge, mass, and spin.

The feedback of the wavefunction of particles within their system of coordinates (i.e., of the wave-transform of the particles' collective quantum state) links them with other particles in their coordinate system. This produces the observed forms of correlation inter alia in the EPR experiment and in the various versions of the split-beam and double-slit experiments. In split-beam experiments with "which-path" detectors the feedback of the system's collective wavefunction makes for the correlation of the photons, electrons, or atoms with each other, as well as with the experimental apparatus. It appears that labeling the particles for detection is sufficient to establish correlation with the apparatus: particles of which the path can be read out by the apparatus are effectively part of the particle-*cum*-apparatus coordinate system, whether or not the path is actually read out.

Particles are in-formed with the wavefunction of their system of coordinates due to the selective matching of their quantum state (through the form of their wavefunction) with the plenum-conveyed collective wavefunction, that is, with the state of the coordinate system. Selection is required, because even if the highest-order wavefunction—that of the universe—is common to all particles in the universe, lower-order wavefunctions differ according to the quantum state of the component particles. Therefore, notwithstanding the in-formation of all particles with the same superordinate wavefunction (the wavefunction of the universe), each particle is effectively entangled only through that component of the universe's wavefunction that corresponds to, i.e., is immediately conjugate with, its own state.

Selectivity of this kind does not call for conscious intelligence: it is familiar from experience with conventional holograms. A given holograph interference-pattern meshes with, and can be used to pick out, a conjugate pattern within a conceivably vast array of diverse patterns.

The microworld hypothesis provides an explanation of the temporal as well as the spatial dimensions of quantum entanglement. The temporal dimension—the quasi-infinite endurance of the entanglement—is accounted for by the endurance of frictionlessly propagating superposed interference patterns in the Ψ field, the scalar field of the plenum. The spatial dimension—the quasi-instantaneity of the entanglement over any known finite distance—is explained in turn by the variable-velocity (and in matter-dense regions supraluminal) propagation of scalar wavefronts in that field.

The Macroworld Hypothesis

As already noted, in-formation through the Ψ field is scale-invariant: it occurs in the microworld of the quantum, in the mesoworld of the organism, as well as in the macroworld of the cosmos. Thanks to the coding and transmission of the respective wavefunctions in this field, charged particles are correlated in their system of coordinates by the wavefunction of their coordinate system, molecular and cellular assemblies are correlated in the organism by the wavefunction of that organism, and planets, stars, stellar systems, and galaxies are correlated by the wavefunction of the universe.

Menos Kafatos, Robert Nadeau, and other cosmologists have shown that nonlocal types of correlation extend throughout space-time. Nonlocality in the universe suggests the presence of the collective wavefunction of the galactic macrostructures, which is the wavefunction of the universe. Here the top segment of reciprocal Ψ field mediated in-formation is relevant:

$$S_{gal} \rightleftarrows \Psi field\ \Psi(S_u) \qquad (9)$$

where S_{gal} stands for galactic and supergalactic structures, and S_u for universe.

The in-formation of cosmic macrostructures with the wavefunction of the universe means that to some subtle but not negligible extent all macrostructures are correlated with each other. This elucidates two major puzzles of contemporary cosmology.

As we have seen in chapter 1, stars and galaxies evolve isomorphically in all directions from Earth. The large-scale uniformity of the macrostructures extends even to galaxies that are not in physical contact with each other since they recede from one another at a rate higher than the speed of light. Mainstream cosmology's answer to this puzzle is the introduction of Guth's and Linde's "period of inflation." During Planck-time (the first 10^{-35} second following the Big Bang) all parts of the universe were in physical contact. This is said to explain the uniformity of cosmic evolution even if the subsequent expansion of the universe had prevented light rays from catching up with the expansion of the periphery and reaching the outer galaxies. The microwave background radiation observed through the balloon-based Boomerang telescope, together with the findings of the MAXIMA and DASI project teams working with a ground-based instrument at the South Pole, identified a primary resonance and two higher-frequency harmonics that could have been due to an inflationary burst in the very early universe.

However, the very same facts can also be explained by a cyclic model of the universe. In the Steinhardt-Turok scenario each cycle in the pulsating metaverse goes through the currently observed period of slow accelerated expansion, followed by a period of contraction that leads to the homogeneity, flatness and the energy required to start a new cycle (Steinhardt & Turok 2002). Not only can this cyclic scenario account for the facts covered by the inflationary scenario, it can also account for observations that are beyond the scope of the standard model. Inflationary Big Bang theory can explain the observed fluctuations of the microwave background together with the distribution of the galaxies and the homogeneity and isotropy of the universe on large scales (>100 megaparsec), but it cannot shed light on the initial conditions that had determined the parameters of the observed universe, and it cannot account for the recent discovery of cosmic acceleration and the self-repulsive dark energy that is presumably responsible for it. As we shall see, Steinhard and Turok's cyclic model can do so.

Independently of these considerations, it is highly doubtful that inflation alone could account for the observed coherence of the macrostructures of the universe. For such coherence to occur and to persist, some form of interconnection among the structures may be required. This is provided by the hypothesis that the scalar wavefronts that encode the wavefunction of the universe, unlike the propagation of light in the electromagnetic field, is supraluminal and thus keeps up with the expansion of the galaxies.

The cosmological result of Ψ field in-formation is the coherent evolution of the galactic macrostructures: $\Psi(S_u)$ in-forms S_{gal}—the wavefunction of the universe interlinks its parts. On the scale of the cosmos, this in-formation correlates the pathways of galactic evolution.

The macroworld hypothesis elucidates also another puzzle of contemporary cosmology: the fine-tuning of the universal constants. We have seen that the values of the constants are fine-tuned to such an extent that the statistical probability that they were hit upon by random selection is only significant if we assume a very large number of universes. Since fluctuations in the pre-space of the universe have determined the parameters of the Big Bang and therewith set the values of the constants in the resulting universe, the decisive question concerns the nature and origin of these fluctuations.

The standard model maintains that pre-space fluctuations were random. This assumption is contradicted, however, by the statistical analysis of the microwave background. It appears that its inhomogeneities come in certain specific patterns, a minute portion of all the patterns that are likely to occur in a random process. And the small subset of patterns that did occur was precisely such that a coherently evolving and ultimately life-bearing universe could come into existence.

The selectivity of the pre-space fluctuations is not meaningfully accounted for by the law of large numbers. Assuming that there are a vast number of universes, and in each universe a small subset of the possible fluctuations occurs by random selection, does provide a statistical probability for the kind of fluctuations that gave rise to our universe, but it does so at the cost of the intrinsically unverifiable (and in the final count unnecessary) assumption of a vast number of universes with independent randomly selected features. In turn, the anthropic principle—which in its various versions claims either that the universe is the way it is because there are conscious observers of it, or at least that it is the way it is because conscious observers can ask questions about it—is ad hoc and frankly anthropocentric.

The most consistent and economic explanation of the fine-tuned features of the universe is that its pre-space was structured by a precursor universe (or series of universes). If so, the fluctuations that tuned the Big Bang—which in turn determined the values of the universal constants—did not result from a chance selection from among an astronomical number of alternatives, but were due to traces of prior universes which limited the plenum's degrees of freedom. This is the minimal "transempirical" assumption that can shed light on the universe's anomalous coherence. It is not unduly speculative: as we have seen, a number of cosmologists postulate a metaverse in the framework of which our universe would have arisen.

How local universes would arise in the metaverse is still debated. One possibility is that the extreme high densities of black holes produce singularities where the known laws of physics do not apply. Under these conditions the black hole's region of space-time detaches itself and expands to create a universe of its own. Another possibility is that "matter-creating events" intersperse the evolution of galaxies. The QSSC (Quasi-Steady State Cosmology) advanced by Fred Hoyle, G. Burbidge, and J. V. Narlikar suggests that matter-creating events come about in the strong gravitational fields associated with dense aggregates of preexisting matter in the nuclei of galaxies (Hoyle et al. 1993). While a superposed oscillation period of forty billion years underlies the universe's overall expansion, the most recent burst is said to have occurred about fourteen billion years ago, in reasonable agreement with the standard model.

The cosmology put forward by Ilya Prigogine, J. Geheniau, E. Gunzig, and P. Nardone agrees with the QSSC in suggesting that major matter-creating bursts similar to the Big Bang occur from time to time (Prigogine et al, 1988). It specifies that the large-scale geometry of space-time creates a reservoir of negative energy (which is the energy required to lift a body away from the direction of its gravitational pull) and from this reservoir gravitating matter extracts positive energy. Gravitation is thus at the root of the ongoing synthesis of matter: it produces a perpetual matter-creating mill. The more particles are generated, the more negative energy is produced, transferred as positive energy to the synthesis of still more particles.

In the cyclic model of the universe advanced by Steinhardt and Turok the universe undergoes an endless sequence of cosmic epochs each of which begins with a "Bang" and ends in a "Crunch." Each cycle includes a period of gradual accelerated expansion followed by contraction, and leads to the conditions of homogeneity, flatness and energy needed to begin the next cycle. Negative potential energy and not spatial curvature accounts for the reversal from expansion to contraction. The continuation of the currently observed accelerated expansion dilutes the entropy, black holes, and other remnants of evolution in the present cycle, returning the universe to a pristine vacuum state where it contracts and bounces to a new cycle. Temperature and density are finite at the transition between the cycles.

In the Steinhardt and Turok model the universe is infinite and flat, rather than finite and closed, as in other cosmologies that postulate a bounce from contraction and crunch to explosion and expansion. This cyclic scenario consists of the evolution of a scalar field Φ along a potential $V(\Phi)$ in the framework of a four-dimensional (4D) quantum field theory. Its essential features are the form of the potential and the coupling between the scalar field, matter, and radiation. At the beginning of a cycle the scalar field Φ is increasing rapidly, but its expansion is damped by the expansion of the resulting universe. Eventually Φ comes to rest in a radiation-dominated phase and remains nearly static until dark energy begins to dominate and acceleration resumes. At present we are about 14 billion years into the current cycle and at the beginning of a trillion-year period of acceleration.

The energy of the universe's ground state is negative, rather than zero. However, between the cycles the universe never reaches the true ground state, hovering above it, bouncing from one side to the other, but spending most of the time on the positive-energy side. The quasi scale-invariant fluctuations during the contracting phase transform into a quasi scale-invariant spectrum of density fluctuations in the phase of expansion. Hence all or nearly all regions of the universe undergo the same sequence of processes, with most of the time in a cycle spent in the radiation, matter, and dark-energy dominated phases (Steinhardt & Turok 2002).

The Steinhard and Turok scenario, as all multicyclic cosmologies, constitutes a complete model of cosmic history, unlike the standard model where inflation and the subsequent acceleration follow an unexplained creation event. But as presently formulated, multicyclic cosmologies do not give an account of the fine-tuned features of the observed universe. They can, however, be developed to do so. This requires to hypothesize that *the pre-space of a new cycle is effectively in-formed by events in the preceding cycle.* Trans-cyclic "in-formation" can obtain if the fundamental scalar field of the cosmos conserves information on the charged particles that emerge and evolve in the various cycles. The presence of this information in the pre-space of a new cycle affects the fluctuations that tune that cycle's basic parameters. In this manner the cycles exhibit a learning curve, leading to progressive consistency among the successively arising universes.

This theoretical advance is anticipated in the macroworld hypothesis. Here the in-formation of the Ψ field (which about 13.7 billion years ago constituted the pre-space of our universe) is given by the formula

$$S_u \rightarrow \lrcorner \ \Psi \mathit{field} \ \Psi(S_{mu}) \qquad (10)$$

where S_u stands for our universe, and S_{mu} for the metaverse. (Evidently, if (10) is true, S , the highest-level system, is S_{mu} rather than S_u.)

In-formation by prior universes (or universe-cycles) occurs through the scalar field of the cosmic plenum, the basic substratum common to all universes (or universe-cycles). Interfering scalar wavefronts code and conserve the wavefunction of the particles and systems of particles that evolve in space and time. The wave-interference patterns diffuse at variable velocity throughout the plenum of the given universe, and since they diffuse frictionlessly, they endure indefinitely over time. Consequently the patterns created in prior universes (or universe-cycles) must be assumed to have been present in the pre-space of our universe. They constrained the randomness of its fluctuations, limiting them to those which set the values of its universal forces and constants consistently with the forces and constants of prior universes.

If the set of prior universes included universes with a significant capacity for complex-system evolution, the fluctuations of the scalar field in the pre-space of our universe tuned its Big Bang so as to create universal laws and constants that permit the evolution of galaxies with stellar systems, and occasional solar systems with planets capable of supporting life.

The Mesoworld Hypothesis

In the mesoscale world of life and mind a dual segment of the reciprocal process of plenum-structuring and consequent particle and particle-system in-formation is involved. Unlike quanta, which do not have subsidiary parts, and the universe (or metaverse), which is not itself part of a larger whole, organisms are wholes in regard to their constituent cells and molecular assemblies, and parts with respect to the ecological and socioecological systems in which they participate.

$$S_{cell} \rightarrowtail \Psi field\ \Psi(S_{org});\ \text{and}$$
$$S_{org} \rightarrowtail \Psi field\ \Psi(S_{eco}) \qquad (11)$$

This two-way process comprises

1. the structuring of the Ψ field with scalar interference patterns that carry the wavefunction of molecular and cellular assemblies (S_{cell}), and the in-formation of the molecular and cellular assemblies with the macroscopic wavefunction of the organism of which they are a part (S_{org}), and

2. the structuring of the Ψ field with the wavepatterns that carry the wavefunction of organisms (S_{org}), and the in-formation of organisms with the wavefunction of the ecological or socioecological systems in which they participate (S_{eco}).

Segment (1) accounts for intra-organic coherence, the anomalous coherence of the organism itself, and segment (2) for transorganic coherence, the coherence of the organism with its milieu.

According to connectivity hypothesis the ground state of the cosmic plenum provides a universal dynamic background for all possible excitations, whether at the microscale, the mesoscale, or the macroscale. At the microscale the modulation of the plenum through quantized excitations carries the wavefunction of charged particles and coordinate systems of particles, while at the mesoscale it carries the macroscopic wavefunction of organisms and systems of organisms. Since living organisms have a quantum-correlated aspect, we can speak of the wavefunction of the organism: a macroscopic wavefunction. Bernd Zeiger and Marco Bischof pointed out that this wavefunction provides a global collective description of the dynamic evolution of living systems in terms of a classical but complex wavefield with a well-defined quantum phase. In this way a biological system is a coherent whole that can be described by a single wavefunction similar to the wavefunction of an atom or a molecule (Zeiger & Bischof 1998).

In-formation with the organism's macroscopic wavefunction accounts for the overall coherence of the organic state. This coherence suggests a multidimensional, quasi-instant correlation among all parts of the organism, beyond the range of known biochemical effect transmission. Organic coherence is maintained even if some parts of the organisms are not contiguous, as shown by Backster's experiments, where in vitro cells removed from the organism and placed at finite distances from it respond the same way as cells do within the organism (cf. chapter 2).

The Grinberg-Zylberbaum stimulus-transfer experiments and the Montecucco wave-pattern synchronization tests show that the brain functions of individuals with personal or emotional ties to each other are significantly correlated; stimuli are transferred and EEG waves are synchronized (cf. chapter 3). Transorganic coherence of this kind is accounted for in turn by the in-formation of the brain of the test subjects with their collective wavefunction.

Analogously to the quantum world, where the plenum's response is the feedback of the collective wavefunction of a system of coordinates to the particles embedded in that system, in the living world the plenum's response is the feedback of the integrated wavefunction of two or more organisms to each of those organisms. As also noted in chapter 3, individuals with close empathies and emotional ties often find that they are transpersonally linked. The here cited transferred-stimuli experiments show that empathetic and emotional contact is a major factor in transferring the stimuli. The brain of people in close personal relationships appears to be enduringly correlated. Their collective wavefunction affects the brain and infuses the mind of each individual, creating the observed correlation between the states of their brain and the associated states of their consciousness.

Not only two or a handful of individuals can be transpersonally connected; entire societies and ecologies can be so linked, even if such linkages are more subtle and less manifest.

Nonetheless, the feedback of the collective wavefunction of a social or ecological system has observable consequences, affecting both genetic mutation and social behavior.

The genetic consequences of in-formation with the wavefunction of an ecology favors mutations consistent with conditions in that ecology: genetic rearrangements become tuned to outcomes that produce functional solutions to the problems of survival in the milieu. This apparent "pre-adaptation" does not imply a teleological agency: it results from a physical process—the correlation of the state of the organism with the state of its ecological niche. Herewith the mesoworld hypothesis provides a plausible explanation of the descent of species from common ancestors within the known timeframes by eliminating the unrestrained randomness of genetic variation postulated in the Darwinian theory.

In complex species the behavioral consequences of in-formation with the wavefunction of the social or ecological system in which the individual participates emerge into prominence. Some patterns of animal behavior that are said to be "inborn" and ascribed to "instinct" may be due not to intrasomatic genetic coding, but to extrasomatic plenum coding. Behavior patterns acquired during generations of trials and errors become part of the wavefunction of the species. The species wavefunction, coded in the Ψ field and accessed by the members of that species, informs behavior by transmitting the adaptive responses to the challenges of survival and reproduction achieved by previous generations. This creates evolutionary learning. Members of a species do not need to reinvent adaptive responses by continuing to engage in the slow and uncertain processes of trial and error. Doing so would have either precluded the evolution of higher species of organisms, or would have led to their rapid extinction.

For example, the fight-or-flight-response of most mammalian species, evident shortly after birth, involves but a few milliseconds during which the animal must choose the direction of escape and mobilize all the energy required for fast motion, all the while maintaining balance and orientation. The pattern develops too soon to be the result of learning by trial and error, and it is too quick and complex to be transmitted entirely by reading genes stored in DNA through RNA, proteins, enzymes, and other biochemical transmitters. Instead, this and similar "inborn" behavior patterns are most likely due to the transmission to the animal's nervous system of the wavefunction of a behavior perfected by previous members of the species through a series of trials and errors subjected to natural selection. The adaptive behavior's wavefunction becomes part of the wavefunction of the species and enters the brain and nervous system of successive generations of individuals.

Because living organisms are in-formed by the collective wavefunction of their species and as a result access the survival-oriented patterns of behavior developed by their progenitors, biological evolution is both Darwinian and Lamarckian: in addition to the genetic mechanism of inheritance, it has an acquired dimension, consisting of the transgenerational transmission of the information required for survival.

On the human level the information conveyed through the Ψ field acquires a psychological dimension. The individual's brain is in-formed with the collective wavefunction of the human species as well as with the wavefunction of the socioecological system in which the individual participates. In modern people this information is repressed from waking consciousness, but remains present in the subconscious domains of the mind. Transpersonal intuitions and images, together with species-wide "archetypes" and archetypal elements are present in the psyche of all normal individuals. They shape the emotional and intuitive aspects of behavior and may surface to conscious awareness in meditative and other altered states. The cerebral activity of generation after generation of individuals, integrated in the collective wavefunction of the species, in-forms the brain and nervous system of successive generations of individuals.

Conclusions

Einstein's pronouncement, "we are seeking for the simplest possible scheme of thought that can tie together the observed facts," is the aspiration as well as inspiration of connectivity hypothesis. This hypothesis seeks to tie together in the simplest possible scheme of thought a particular kind of observed fact: that which pertains to coherence resulting from intrinsic, time- and distance-independent correlation in areas as diverse as the quantum, the cosmos, and life and consciousness. For this hypothesis coherence is the explicandum, and active "in-formation" the explicans.

The presence of in-formation in nature is a fundamental tenet of connectivity hypothesis. The hypothesis rests on three general propositions—three "laws" of in-formation as a physically active element in the universe:

(i) charged particles and systems constituted of charged particles create physically active information;

(ii) the information is conserved;

(iii) the information created and conserved feeds back to ("informs") charged particles and systems of particles.

Connectivity hypothesis shows that the feedback of active in-formation, occurring in the holographic mode, creates coherence among the particles and systems of particles that created it.

In the realistic perspective adopted by this hypothesis, the conservation and feedback of in-formation presupposes a physical medium that is best conceived as an extended universal field. Such a field is not substantially different from other fields postulated in science. General relativity's G-field, for example, explains phenomena as diverse as the falling of apples to the ground, the movement of the pendulum and the trajectory of the planets around the Sun, as well as some phenomena that are unobservable, such as the curvature of space-time. The G-field itself is not an observable, yet it is a key element in what Einstein and generations of physicists considered the simplest consistent scheme of thought that can account for the observed facts.

Whether postulating a cosmically extended information-conserving field gives us the simplest consistent scheme that can explain the facts of anomalous coherence can only be ascertained on the basis of sustained testing in the relevant fields of scientific interest. We can already affirm, however, that in light of the finding of space- and time-independent coherence in physics, cosmology, the life sciences, and consciousness research, there are few if any realistic alternatives to postulating such a field in nature. Connectivity hypothesis specifies that this field is constituted of wave-interference patterns generated by charged particles in the scalar field of what has been misleading termed quantum vacuum. This specification is cogent, since it is now recognized that

1. wave-interference patterns are uniquely qualified as mechanisms for coding, storing, and transmitting large quantities of information;
2. all presently known fundamental fields are rooted in the "unified vacuum," the cosmic plenum; and
3. a scalar field is known to be associated with the electrostatic potentials of the plenum's zero-point field.

Just one additional assumption is required to account for the full range of anomalous coherence in nature, from the microdomain of the quantum through the mesodomains of life and mind, to the macrodomain of the universe.

4. Much as the vectorial fields of the cosmic plenum conserve and convey *energy*, so its scalar field conserves and conveys *information*. This nonvectorial "in-formation" correlates parts within wholes, and wholes within sequentially higher-order wholes and creates coherence within and among them.

Chapter 9

♦

The Advent of Integral Quantum Science

Historians and sociologists of science often remark that scientific knowledge grows not only, or even primarily, through the sustained accumulation of observations built into preexisting theories, but through leaps from one fundamental theoretical conception to another. Such paradigm-shifts, termed scientific revolutions, occur periodically in the course of science's development.

In the period from the seventeenth to the nineteenth century science was in rapid yet relatively linear evolution. It built on the paradigm provided by Galileo, Kepler, and Newton and, emancipating itself from religion gained a dominant position in the Western world. The twentieth century witnessed a number of revolutions, first in physics, and subsequently also in biology, cosmology, and consciousness research. Science's impact in society grew, mainly through physics-based breakthroughs in transportation, production, information-processing, and communication. However, toward the end of this period researchers in fundamental physics, cosmology, biology, and consciousness studies encountered deepening anomalies. Now another leap is about to occur—another scientific revolution.

The outcome of the coming revolution is variously assessed. A number of observers believe that, given current advances in genetics and the spread of an organic approach to natural as well as human ecology, the twenty-first century will be a century of biology. This view has much to recommend it, but it does not grasp the full range of the anomalies that drive the current development. In the opening years of the twenty-first century the evolution of science is driven by the discovery of space- and time-invariant coherence not only in quantum physics, but also in biology, cosmology, and consciousness research. Quantum physics gives a sophisticated mathematical account of quantum coherence (although if fails to give a realistic explanation of it), but in most other fields the analogous forms of coherence are mainly anomalous. Space- and time-invariant coherence in the diverse domains of investigation conflicts with the paradigm of local action and localized causality that dominates the biological and human sciences.

The finding of enduring, instantaneous coherence in phenomena is not just a paradox; it is also a spur for theory-innovation. As a quasi-universal phenomenon it requires a new conceptual framework, one that can exhibit the unity of the main branches of the empirical sciences including physics, cosmology, biology, and the transpersonal and quantum brain-theoretical schools of consciousness research.

Coherence, of course, is not the only factor arguing for the unity of the physical, the biological, and the psychological sciences. Despite important differences at the level of observation, on deeper analysis significant continuities are coming to light among the phenomena investigated in these sciences. Evolution in the universe and evolution on Earth, though phenomenologically different, prove to be continuous and in some respects mutually consistent. There is, for example, a continuous and consistent buildup of free energy density in physical and biological systems. Eric Chaisson has shown that F_m, the value of free energy rate density (the unit of energy per time per mass, erg s^{-1} g^{-1}) increases throughout the range of physical and biological evolution. For stars the average value of F_m is 2; for planets such as the Earth it is 75; for plants in the biosphere it is 900; and in the human body it is 20,000 (Chaisson 2000).

Beyond free energy density a wide variety of physical-biological invariances have been investigated by such "transdisciplinary disciplines" as cybernetics, information theory, and general systems and general evolution theory (Laszlo 1987). Building on these continuities and invariances, science is currently growing beyond physics and beyond biology, into the dimension of transdisciplinary theory.

The rise of an integral science of truly transdisciplinary scope is radically new, it constitutes another scientific revolution. The hitherto advanced unified and grand-unified theories, and the string and the related "theories of everything" are mono-disciplinary: they are unified theories of *physics*, at best theories of every *physical* thing. By contrast the integral science now on the horizon promises to be a science of physical as well as of biological, and even of psychological "things." It will embrace quantum physics and quantum biology, as well as quantum cosmology and quantum brain and consciousness research. It will be a transdisciplinary field of research and experimentation applying concepts developed in the microscopic domain across the full range of observed phenomena.

BASIC CONCEPTS—Two concepts will function as root metaphors of the integral quantum science of the 21st century: *fields* and *information*.

The sustained investigation of the cosmic plenum as the basis of the entire realm of manifest phenomena, including mass, energy, and information, will highlight the role of fields not only in physics and cosmology, but throughout the range of observed phenomena, including the phenomenon of mind. The reinterpretation of general relativity's geometric space-time as the locus of a universal field that not only gives rise to matter-energy entities and systems, but also links them and conserves their traces, will shed light on the phenomenon of anomalous coherence and will build it into the scientific world picture. A sound hypothesis of connectivity will lay the foundation for a science that is more inclusive, and penetrates deeper into the realms of reality than the mainstream sciences of the twentieth century. The integral quantum science of the twenty-first century will offer a realistic mapping of the pertinent facts, regardless of whether they pertain to the physical aspect of reality, or to its biological or psychological aspect.

Information will be the second root concept of integral quantum science. Information is not only the dominant reality of technological civilization; it is also emerging as a basic feature of the investigation of nature. According to Roy Frieden, it is the foundation of the laws of physics (Frieden 2001). His work demonstrates that the laws that govern the physical world are derivable from the amount of information present in observed phenomena. Frieden points out that the much vaunted equations of quantum physics, considered the most basic laws of the known world, derive their legitimacy from the fact that they work: they have been tested over and over again, with a number of predictions confirmed up to ten places of decimals. However, today's quantum theories do not disclose why the laws take the form they do. Frieden finds that the form of the laws can be derived by applying I, so-called Fisher information (the formula for determining how much information one can obtain from a physical system) to J, the amount of information bound up in the system being measured. Both I and J can be calculated for a wide range of phenomena. To derive a law of physics (more exactly, the Lagrangian that defines that law) we need to define the precise location of the system in space and time and subtract J from I. This leads to the appropriate Lagrangian, and when it is made as small as possible, the pertinent law of physics emerges. Information, Frieden maintains, is what physics is all about.

Information is what all empirical science, and not just physics, is all about, yet the origins and status of the information discovered in nature remain to be clarified. Following Wheeler's suggestion, that observer participancy gives rise to information, and information gives rise to physics, Frieden speculates that the quantity of information inherent in a system under observation is created in the act of observation. However, information may also be objectively present in nature. If so, the act of observation does not *create* the information we find, but merely *elicits* it.

Integral quantum science will recognize that information not only defines the form taken by the laws of nature, but is a physical factor that connects phenomena and informs their behavior. Information in this sense is "in-formation": the nonenergetic "formation" of the recipient by the message.

In the 1950s David Bohm's hidden variable theory contained an explicit—if still classical—concept of in-formation: the "quantum potential." A complex factor that reflects the entire context of quantum measurements, the quantum potential guides the path of the electron and allows a causal interpretation of quantum phenomena. Though a classical factor, the quantum potential was said to act by form alone, and hence it anticipated the notion of physically active in-formation. The latter Bohm developed in the late 1980s in his "ontological interpretation of quantum theory." Here quantum processes—the processes by which a determinate physical outcome emerges out of a multiplicity of potentialities—are accounted for in reference to a holograph field that produces active in-formation (Bohm 1980, Bohm & Hiley 1993).

This concept had precedents throughout the twentieth century. Einstein's own concept of the Führungsfeld (guidance field) mentioned by him in the 1920s was basically a nonenergetically in-forming field, governing the motion of particles in space-time. Although Einstein came close to incorporating this concept in his subsequent unified field theory, he did not develop it in theoretical form. In general relativity he opted instead for the geometry of spacetime to guide the motion of particles—possibly because too little was known at the time about the quantum vacuum to permit the assumption that it would constitute a physical field capable of affecting the behavior of charged particles. Einstein did, however, note that the concept of a "physically real ether" must be reintroduced into the worldview of physics, and his insight is now gaining validity. Bold new theories interpret the equations of general relativity as equations of motion in a physically real universal substratum—the "physical ether"—instead of as equations that define the formal geometry of space-time (cf. chapter 5 and Appendix 1).

Although Bohm did not generalize the concept of nonenergetic in-formation beyond physics, it is now evident that physically effective yet nonvectorially propagating in-formation is not limited to the quantum world. Evidence reviewed in this study shows that it is a factor in the evolution of the living world, of the world of consciousness, and of the universe as a whole.

Presently Harold Puthoff, Roger Penrose, Glenn Rein, A. E. Akimov, Fritz-Albert Popp, László Gazdag, Hans Primas, Marco

Bischof, and other front-line investigators follow up Bohm's and Einstein's insight and explore the fundamental role of fields and in-formation in a wide range of phenomena of scientific interest. Puthoff articulated the basic insight and the challenge it poses to science: "... a dynamic equilibrium exists between the ever-agitated motion of matter on the quantum level and the surrounding zero-point energy field... Who is to say whether... modulation of such fields might not carry meaningful information?" If this research comes to full fruition, he added, "what would emerge would be an increased understanding that all of us are immersed, both as living and physical beings, in an overall interpenetrating and interdependent field in ecological balance with the cosmos as a whole, and that even the boundary lines between the physical and "metaphysical" would dissolve into a unitary viewpoint of the universe as a fluid, changing, energetic/informational cosmological unity" (Puthoff 2001).

* * * * *

Built on the foundation of a fully developed and consistently tested hypothesis of connectivity, integral quantum science will penetrate deeper into the domains of reality than the physical, biological and psychological sciences of the twentieth century—below the level of the quanta that populate space-time, to the cosmic plenum that generates the quanta and interconnects the particles and systems built of them. It will also penetrate wider into the cosmos—beyond the spatial and temporal boundaries of this universe, to the metaverse that gave birth to this universe and set its parameters. These extensions of the penetration of science will not be arbitrary, or even surprising: they are the logical continuation of the series of conceptual breakthroughs that extended scientific inquiry from the sphere of immediate observation to the wider and deeper domains of instrumental observation, carried ever further by conceptual analysis and mathematical extrapolation.

In the seventeenth century Newton's classical mechanics gave us the mechanistic universe, with independent mass points externally connected by deterministic causal relations. In the twentieth century Einstein's general relativity gave us the relativistically interlinked universe, where all things are connected by signals propagating across the geometric structure of space-time. In the twentyfirst century integral quantum science will give us the coherent universe, where all things are intrinsically connected by subtle yet effective in-formation conveyed by a fundamental virtual-energy field at the heart of a possibly infinite metaverse.

Postscript

◆

The Metaphysics of Connectivity

Metaphysics, according to Aristotle, follows physics as the study of the first principles entailed by our understanding of the nature of reality. Every conception of the nature of reality entails a metaphysics, whether it is recognized or not. Conscious recognition of the indicated metaphysics is both intellectually satisfying—it spells out the conceptions that underlie our theories—and is important in bringing to the surface half-acknowledged assumptions that nonetheless guide observation, experimentation, and the interpretation of the findings.

We begin the elucidation of the metaphysics suggested by the finding of quasi-universal space- and time-invariant connectivity in nature with the biggest question of all: the fundamental shape of reality.

Two Domains of Reality

Universal connectivity makes reality unitary but it does not make it uniform: it can be analytically segmented into two principal domains. One is the manifest domain of (directly or instrumentally) observable particles and systems of particles; the other the virtual domain of the cosmic plenum, the energy sea from which the particles arise, with which they interact, and into which they ultimately fall back. The latter domain is intrinsically unobservable, but it is inferable through its effects on the observable domain. We speak of the former as the "manifest domain," and of the latter as the "virtual domain." (In this context "virtual" is not opposite to "real," but to "observable.") The interaction of the two domains generates the observable entities—the particles and systems of particles—of the universe. The virtual energy-domain is both the generative ground of these entities, and the medium that correlates their state and informs their evolution.

Quantized particles and the systems built of them are the furnishings of the manifest domain. Their apparent materiality does not represent derivation from a categorically disjunctive element of physical reality denoted "matter." Quantized particles and the systems constituted of them are force-like, light-like, or (if endowed with rest-mass) matter-like, but in their ontological reality they are vibrating nodal points (distillations or crystallizations) of the energies of the virtual domain. Virtual energies become manifest—that is, emerge from the virtual into the manifest domain—upon the "excitation" of the virtual domain by an intrinsic instability (as in the universe-creating explosion known as the Big Bang) or by the influx of a significant level of energy (as in ordinary pair-creation). They are sustained in the manifest domain by interaction with each other. In the absence of interaction, the particles do not exhibit the corpuscular properties that hallmark the manifest domain: they remain part of the underlying virtual domain without unique location in space and time.

At various levels of evolution the manifest entities of the universe are particles, systems of particles, and higher-order systems of systems of particles. They are similar to Alfred North Whitehead's "actual entities" and "societies of actual entities" (or generally "organisms") (Whitehead 1929, 1976). They bind the energies emerging from the virtual domain in quantized packets. Charged particles and the systems formed of them are internally related to each other and to the rest of the universe. They are what they are because they receive (a) vectorial-energy signals from other particles and particle-systems in their surroundings, and (b) nonvectorial "in-formation" regarding the state of particles and systems of particles throughout the universe. The wave interference-patterns that convey the latter are the formative elements in the evolution of complexity in the cosmos: they are the physically real counterparts of Plato's "Forms" and Whitehead's "eternal objects." They are not given a priori, since they are the result of interaction between the universe's virtual domain and its manifest domain.

Thus in regard to the formative interference-patterns of the cosmos, the metaphysics of universal connectivity departs from Whitehead's metaphysics, and also from the philosophy of Plato, which was Whitehead's inspiration. The a priori given of the universe is not a set of formative patterns, but a two-fold potentiality residing in the virtual domain. It is the potentiality of that domain (1) to create manifest particles, and (2) to encode the wave-patterns that correspond to the state of the particles and of the systems built of them, and to convey the patterns to in-form the state of other particles and systems of particles.

In the metaphysics of universal connectivity the ultimate reality is the virtual-energy domain from which spring manifest particles and systems of particles in space and time. The two domains, the emergent domain of manifest entities and the fundamental domain of structured virtual energies, are the universe in a complete envisagement. They are categorically distinct but not ontologically disparate: they are fused through the cyclic processes of evolution and devolution.

The two domains of the universe are diachronically as well as synchronically related. Diachronically, the virtual domain is prior: it is the generative ground of the particles and systems of particles that populate the manifest domain. Synchronically, the generated particles are linked with their generative ground through an ongoing bi-directional interaction. In one direction the manifest particles structure the virtual energy domain in which they are particularized nodal points or crystallizations. In the other direction the structured virtual energy domain in-forms the manifest entities that subsist and evolve in space and time. In-formation leads to the coherent evolution of the manifest domain, which in turn further structures the virtual domain.

This conception corresponds to a perennial intuition also articulated in Hindu cosmology. There the almost infinitely varied things and forms of the manifest world are united in an essential oneness at a deeper level. At the fundamental level of reality the forms of existing things dissolve into formlessness, living organisms exist in a state of pure potentiality, and dynamic functions condense into static stillness. All attributes of the manifest world merge into a state beyond attributes. Time, space, and causality are transcendend in a state of pure being: the state of Brahman. Absolute reality is the reality of Brahman; the manifest world enjoys but a derived, secondary reality—mistaking it for the real is the illusion of maya. Brahman, though undifferentiated, is dynamic and creative. From its ultimate "being" comes the temporary "becoming" of the manifest world, with its attributes, functions, and relationships. The samsara of being-to-becoming, and again of becoming-to-being, is the *lila* of Brahman: its play of ceaseless creation and disssolution. The absolute reality of Brahman and the derived reality of the manifest world constitute an interconnected whole: the advaitavāda of the universe.

The Interactive Evolution of the Virtual and the Manifest Domains

At the logically extrapolated (but empirically unverifiable) beginning of the cosmic process only the virtual domain existed, in a primordial state. This was a spatially and temporally unbounded sea of fluctuating virtual energies: the pre-space of the manifest domain. When a region of this cosmic plenum manifested a critical instability, some fraction of the thereby liberated energies became established as quantized packets of matter-energy. The relations of these entities to each other and to the underlying virtual domain launched the evolutionary process with the known dimensions of space and time.

The virtual domain first generated the particles that are the initial and basic constituents of the manifest domain, and then, in a progressive but intermittent and nonlinear evolutionary process, created sequentially more complex multiparticle systems. During the 13.7 billion years that elapsed since the explosive instability that created the universe we now observe, the two domains evolved in reciprocal interaction. As a result the manifest entities of this universe are coherent wholes, and are coherently related to the superordinate and likewise coherent wholes that make up their environment.

The evolution of the universe is "in-formed" in that the wave-transform of the state of manifest entities—their wavefunction—enters the virtual domain and is conserved in it. In sequentially superposing wave interference-patterns the wavefunction of individual particles and systems of particles is integrated in the collective wavefunction of their superordinate systems. The virtual domain encodes the collective wavefunction of the particles and systems of particles. The highest-order wavefunction coded in the virtual domain is the wavefunction of the highest-level system: the wavefunction of the universe. This wavefunction in-forms galaxies and all lower-level matter-energy systems throughout space and time.

The interaction of the manifest and the virtual domains introduces an element of "soft" determinism into the in-formed evolution of the universe. The classical varieties of "hard" determinism apply principally to the relations of parts within a system: these constitute "upward causation" by jointly codetermining the structure and function of the system formed by them. However, in our interactively evolving universe a more subtle yet equally effective form exists as well: the soft-determinism that comes about through the interaction of manifest entities with the formative patterns of the virtual domain. By in-forming the particles and systems of particles with the wavefunction of their superordinate systems, this interaction produces "downward causation." In the quantum world downward causation produces an entanglement of particles within their system of coordinates, in the living world it creates intra- and trans-organic coherence in and among organisms, societies, and ecologies, and in the universe at large it correlates the structure and evolution of galaxies and supergalactic clusters.

Cosmic Evolution

In-formed evolution characterizes our spatially and temporally finite though unbounded universe. Spatial finiteness is due to the finite expansion of galaxies and supergalactic clusters, while temporal finiteness is the consequence of the finite availability of the free energies required for the evolution of manifest entities.

In the course of cosmic aeons irreversible processes exhaust the concentration of free energies, the sources and stores of negative entropy. When entropy overtakes the complexity-buildup process, the evolution of manifest entities reverses. Complex structures break down, yielding to simpler ones, which break down in turn until, near the end of cosmic evolution, atomic nuclei, stripped of electron shells, become supercompacted in black holes. In the final "evaporation" of black holes, the degenerate quantal remnants die back into the energy sea of the virtual domain.

A single-cycle universe comes to eternal rest. But further instabilities in the virtual energy domain may occur, and some of these may be potent enough to create new universes. The thesis of a metaverse giving rise to local universes is cogent, and is embraced in the metaphysics of universal coherence. On this thesis either periodic instabilities in the cosmic plenum produce local universes, or the metaverse itself undergoes cyclic renewal, so that evolution in the universes repeats time after time. However, since in each of the universes (or universe-cycles) manifest entities structure the virtual domain, and that domain in-forms the birth of the successive universes, the evolutionary process does not repeat in the same way. The physical laws and constants of later universes become progressively tuned to conditions achieved in their predecessors.

The in-formed evolution of each universe's manifest domain with a virtual domain structured by prior universes creates a learning curve. Subsequently created universes reach the apex of evolution achieved in prior universes in a shorter span of time, or reach a higher apex in an equal time-span. This learning curve defines the evolution of the metaverse across the cycle of local universes.

The evolution of local universes through cycles of the metaverse introduces a direction into the evolutionary process without a preconceived goal or telos. The process drives toward the progressive structuring of the virtual domain by each universe, and the corresponding in-formation of each universe's manifest domain by the virtual domain. This process is cyclic as regards the manifest domain: quantized particles spring from the excitation of the virtual domain, and at the end of their evolutionary cycle die back into the virtual domain. For the virtual domain, however, the process is linearly constructive: the formative patterns created in that domain by the evolution of systems of particles remain conserved from one universe to another, and accumulate throughout the cycle of universes.

The overall process of evolution in the cosmos can be described in reference to the two fundamental elements *energy* and *information*. In the manifest domain energy is conserved, transformed, and in each local universe becomes progressively unavailable. However, it is recycled in each new cycle of the metaverse. In the virtual domain, which is common to all local universes, information is created and conserved, and it in-forms the manifest domain of each local universe. As a result local universes become both more entropic and more in-formed, and the metaverse, energetically self-recycling, becomes progressively more in-formed.

Philosophical Implications

Universal (or quasi-universal) connectivity in the cosmos harbors implications for issues of fundamental philosophical interest. Here we choose three issues for discussion: materialism versus idealism, freedom and morality, and the concept of the divine and its relation to humans as well as to nature.

Materialism or Idealism—or Matter/Mind Complementarity

Whether the universe is primarily material or essentially ideal is a perennial subject of philosophical debate. The classical, categorically monistic alternatives are the following:

1. The manifest particles that had sprung from the excitation of the virtual energy domain are the basic elements of reality. If so, consciousness in the universe is an epiphenomenon, a local and temporary by-product of the evolution of some species of systems constituted of these particles.

2. The manifest entities that populate space and time are a secondary phenomenon: local and temporary carriers of

the evolving consciousness that constitutes the basic reality. The essential feature of the universe is the evolution of the consciousness that is present in it.

Neither the materialist nor the idealist interpretation lacks experiential support. Scientists who take the position of external observers find only matter-like particles and systems of particles and the forces and fields that surround them: for them the universe consists exclusively of these elements. They can conclude that reality is material—mind and consciousness are epiphenomena. Introspective subjects, in turn, find exclusively perceptions, volitions, feelings, and intuitions, their entire stream of experience is made up of these elements of consciousness. For them all of reality is in the form of mind; the rest is a human construction of conscious experience.

A categorically monistic position espouses one of these viewpoints to the exclusion of the other. This is not the only option. The metaphysics of universal connectivity is ontologically unitary but not categorically monistic: in it both *psyche* and *physis* are defining features. Such a conception is not classically dualistic, for matter and mind are viewed as defining, but not as disjunctive, features—they are complementary aspects of the same evolving reality. These aspects are universal: in the interactively evolving universe matter is not limited to particles, and mind is not limited to organisms. Physical reality evolves into all of reality, and mind is an element throughout evolving reality. The universe is "bipolar": *matter* (in the form of matter-like bound-energy entities) and *mind* (as manifested in the stream of lived experience), are distinct but complementary aspects.

The principle of complementarity is borrowed from quantum physics. Nils Bohr suggested that the wave- and the corpuscular-aspects of a particle are complementary—whether the one or the other comes to the fore depends on the kind of questions one asks and the kind of observations one makes. Independently of whether or not the complementarity principle fully accounts for the properties of the quantum, an analogous principle offers an adequate account of the physical and the mental properties of the universe. Complementarity in this regard means that, whether the physical or the mental aspect emerges for an observer depends on the viewpoint assumed by that observer. In the perspective of the external observer, it is the physical aspect that emerges: even the brain of the observer, seen from the "outside," is a system of neurons embedded in gray matter. In the perspective of introspection, on the other hand, it is the mental aspect that appears: not only the observer herself, but the widest reaches of the cosmos are experienced as elements of consciousness, only interpreted as elements of physical reality. The potentials for both aspects are objectively given: they are encoded in the primordial cosmic plenum, which was the pre-space of our universe, as it was the pre-space of all possible universes in the metaverse. Evolution in the successive universes realizes these potentials through an ongoing interaction between the virtual domain, which is the cosmic plenum, and the manifest domain of the particles and particle systems that populate the space and time of the given universe.

The mental potential is realized in the manifest domain corresponding to the level of evolution attained in multi-particle systems. A comparatively evolved system, such as the human, has a comparatively evolved brain and thus a correspondingly articulated mental potential. This endows the human brain with a highly evolved capacity for receiving sensory signals from the manifest domain, and nonsensory in-formation from the virtual domain. In regard to the latter, the brain is generally in-formed by the wavefunction of the universe and specifically in-formed by the wavefunction of the social and ecological systems in which the individual participates. Sensory information constitutes the familiar contents of everyday experience, whereas nonsensory in-formation, in modern societies generally repressed, comes to light mainly in the form of intuitions, images, archetypes, and the seemingly anomalous contents of altered-state experience.

Freedom and Morality

FREEDOM—The interactively evolving universe is not deterministic; in it complex multi-particle systems have nonnegligible degrees of freedom. This freedom derives from the reception of the rest of the universe in and by the systems, and from the response of the systems to the rest of the universe.

A given system's degree of freedom is determined by its level of evolution. Systems select the information they receive from their surroundings through their sensitivity to ambient fields and forces. This sensitivity is limited in simple systems, but it is highly evolved in complex organisms such as the human.

In complex systems with subhuman perceptive capabilities the selection of external stimuli is not significantly infused with conscious awareness. By contrast in humans a high level of consciousness constitutes both a sophisticated instrument of perception and cognition, and a filter that selects some varieties of percepts and intuitions and blocks others. The formative patterns that ingress in the brain through its interaction with the virtual domain are especially subject to repression: modern people repress the greater part of the nonsensory information available to them.

Perceptual and cognitive capacity constitute one aspect of the freedom of complex systems: the aspect of internally guided selectivity in regard to the reception of the manifest and the virtual domains by the organism. The second aspect of freedom concerns the coupling of reception and action. Comparatively simple systems, such as atoms, molecules and the more primitive forms of life, react to external impulses through physical or chemical reactions and reflexive responses. In evolved organisms, on the other hand, a significant range of "intervening variables" qualifies the coupling between stimulus and response. These shift many of the determinants of behavior from outside the systems to within them.

At the human level the internal determinants of behavior are highly evolved. Some of the determinants are subconscious, such as tacit preferences, cultural predispositions, and a range of unconsciously held prejudices and values. There are also conscious determinants, such as purposive goals and preferences, and consciously held values and beliefs. Conscious selection embraces a set of behavioral alternatives in response to a perception or intuition that ranges from inaction to a variety of courses of action. As a result humans, more than members of any other species known to us, enjoy a twofold freedom in the universe: self-guided selectivity in perception, and self-guided selectivity in behavior.

MORALITY—Human beings can attain a significant level of autonomy within the networks of interaction that embed them:

they can decide their level of participation in the environing systems, and hence their level of coherence with them. Unlike nonhuman systems, human beings can engage in behavior that is incoherent with their social and ecological environment. Such behavior may lead to the reduction of coherence in the environing social and ecological systems themselves. It may impair human communities and local ecosystems and, in view of the ties of interdependence that span the globe, may impact negatively on the socioecological system of the planet in its totality.

Coherence reduction in the human sphere is due above all to the repression of the formative patterns accessible to the brain from the virtual domain. Subtle intuitions of ties between the subject and the rest of the world may be ignored, their very existence denied. The external world appears then as a domain of "nothing but" discrete and unconnected individuals and separate material objects.

Since human self-excemption from the networks of coherence in the universe affects not only the given individual but also other human and nonhuman forms of life, it has moral implications. Physical health requires coherence within oneself—the integrated functioning of body and mind—and mental health calls for coherence between oneself and one's social and ecological environment. The latter aspect affects other people as well as the natural environment: hence it has a moral dimension.

Healthy and moral behavior presupposes openness not only to the full range of information reaching the individual through the senses, but openness also to the subtle intuition of interpersonal, social, and ecological ties that in-form the subconscious or conscious mind, and the willingness to adopt the coherence-enhancing behaviors that are consistent with them.

Two Concepts of the Divine: Theological Perspectives

Two basic concepts of the nature of the Divine and its relation to the world can be developed on the basis of universal connectivity: the minimum and the maximum concept.

The minimum concept invokes the necessity of assuming the work of divine agency in relation to the origins of the metaverse and the selection of its evolutionary path. In this concept God created, not the universe as we observe it, but the potentials for the self-creation of the universe, more exactly, the creative potentials of the metaverse in the womb of which our universe emerged. The alternative to the envisagement of this primordial act is to persist in the belief that the principal laws and constants of nature were randomly selected. This, however, would be contrary to all reasonable estimates of probability. As already remarked, hitting on the observed fine-tuning of the universe's laws and constants by random selection among the possible alternatives is astronomically improbable—yet precise selection must have been the case, because complex systems could not have evolved in the known time frames in its absence.

If not arising by random selection, the laws and constants that define the evolutionary path of our universe must have been determined by constraints left by prior universes in the cosmic plenum that served as its pre-space. This alternative does not apply to the hypothetical yet logically necessary first universe to have been born in the metaverse. If the womb of that initial universe was not entirely randomly structured—which would face the paradox of staggering serendipity—it must have been structured by a transcendent agency. Although the creative act that endowed the metaverse with its self-evolving potentials is not open to empirical confirmation or disconfirmation, it is implied by the ensemble of empirically known facts. It is the minimum supposition required for a cogent account of those facts.

Beyond the initiation of the evolutionary process in the metaverse, the minimum concept does not require supranatural agency. The tenet of an initial creative act without continuing divine intervention is known to theologians as Deism. For the most part they reject it in view of the doctrines of the major religions, all or most of which speak to the actuality, or at least the possibility, of ongoing divine intervention.

Science-minded theologians search for the kind of interventions that would be consistent with the laws of nature. Nancey Murphy, Arthur Peacock, and John Polkinghorne, for example, postulate a "top-down" as well as a "bottom-up" form of divine intervention: the former through the divine provision of an ongoing stream of "information" that shapes the course of events at both microscopic and macroscopic levels, and the latter by shifting the probabilities of otherwise random quantum events on the assumption that variations at that level do not cancel out but produce amplified effects on macroscopic levels (Clayton 1997). Through such interventions God is said to influence the course of events without affecting the laws of nature.

Although this possibility merits attention, straightforward Deism satisfies best Occam's Razor: it is the simplest rationally conceivable account of the relation of the Divine to the cosmos. In this account the information that guides the evolutionary process of this universe, and of the entire cycle of universes in the metaverse, is generated in the process itself. Having been endowed with the potentials for self-creation, the two domains of the cosmos, the virtual and the manifest, evolve each other. The information generated in this process is effective on all levels, from the quantum to the universe. It guides the unfolding of the evolutionary process not as a result of external agency, but as a factor internal to it. Thus, even if God is not required to intervene in the evolution of the universe, it is required by this evolution as its original creator, and the designer of its path of unfolding.

We now consider the maximum concept of the Divine and its relation to the world. This concept can be framed in terms of the Whiteheadian notions of the "primordial" and the "consequent" nature of God.

Evolution, we should note, realizes a twofold potential in the cosmos: a physical potential for the progressive, although intermittent and nonlinear, complexification of manifest entities; and a mental potential for the intermittent yet progressive evolution of consciousness. These potentials were encoded in the primordial virtual-energy domain. In the maximum concept that domain constitutes *the primordial nature of God.*

Upon the termination of the evolutionary process—following the "evaporation" of the last remnants of supergalactic structures in the space and time of the last universe—the potentials encoded in the primordial virtual domain achieve final realization. For the maximum concept the completely in-formed virtual-energy domain constitutes *the consequent nature of God*.

At the initiation of the cosmic process God created the potentials for the in-formed coevolution of the metaverse's cycle of universes. And at the end of the process God contemplates the final fruit and complete record of this cyclically recurring but overall irreversible evolution.

The maximum concept is likewise consistent with universal connectivity. The universe's physical and mental aspects, as already noted, are observer-dependent. For an external observer "the rest of the world" appears in the guise of physical reality; for the introspective observer the whole world, including the observer, appears as part of the stream of mental experience. For manifest entities who observe the world from within, both aspects are available. They can introspect on their own experience, and in regard to the rest of the world they can take the stance of an external observer. On the other hand for a Being who is immanent in the universe the perspective of the external observer is not available: in its introspection the entire world and all things in it appear as elements of mind.

When manifest entities assume the standpoint of an external observer, they find that evolution leads from a field of virtual energies endowed with pure potentials (the primordial cosmic plenum), to a virtual energy-field containing the full record of the realization of those potentials (the fully in-formed Ψ field at the termination of the evolutionary process). And when manifest entities take the stance of the introspective observer, they find that evolution leads from a mystical nothingness filled with potential, to a mystical coherence that embraces everything in the whole of experience.

In the maximum concept the introspective perspective on the scale of the cosmos as a whole is the perspective of God. In this perspective all parts of the cosmos are available to introspection at the same time, and they are available at all times. The evolutionary process occurs within God's consciousness. It leads from an initially unruffled divine consciousness informed by pure potentials, through the spatiotemporal excitation of this consciousness, to a finally unruffled consciousness in-formed with the full record of the realization of those potentials.

The evolution of the metaverse through the cyclic evolution of universes conduces to the full realization of the evolutionary potentials encoded in the primordial cosmic plenum—to the complete coherence of all things that exist in space and time. It marks the full achievement of divine creativity: ultimate coherence in the mind of God.

Appendix 1

✦

General Relativity and the Physical Vacuum

Reconsidering Einstein's Equations in Relation to Connectivity Hypothesis

László Gazdag

In Albert Einstein's general theory of relativity space-time, a geometrical concept, is the source of the gravitational effect. In the theory of the Ψ field, on the other hand, space-time is considered in the form of the "physical vacuum," a physically real energy field. Consequently in Ψ field theory gravitation, the same as all other universal interactions, is traced to the interaction of particles with this field. The formalisms presented in this Appendix demonstrate that the interpretation that led Einstein to view the general relativity equations as referring to geometric space-time can be cogently reinterpreted as equations of motion referring to the physical vacuum.

László Gazdag is Professor in the Science University of Pécs, Hungary, and the author of *Beyond the Theory of Relativity* (Robottechnika Kft. Budapest 1998) and other works (cf. p. 133).

Appendix 1

Preliminary Considerations

Einstein published his general theory of relativity in 1916, and this theory is also his theory of gravitation. The idea that the structure of space-time is determined by mass associated with matter was revolutionary in the early 20th century and has since become a pillar of modern physics. Nonetheless, it is one-sided. Of the four basic interactions why should gravitation alone determine the structure of space-time? According to Einstein there is no absolute space and time: all events have their own space-time structure. But what should we understand under "space-time structure"? The standard conception is that this is the structure that results from some factor that constraints the space-time path of a particle. However, the path of an electron in an electric or a magnetic field is more likely to be constrained by the strength of the electric or the magnetic field than by the strength of the gravitational field. Hence the space-time structure of an electron should not be viewed as determined by gravitation alone; we must consider that it may also be determined by the electromagnetic field.

If this proposition is cogent, the equations of general relativity require modification. We can speak of gravitational space-time, determined by the gravitational field, and can also speak of electromagnetic space-time, the structure of which is determined by electromagnetic bosons and antibosons. In the case of massive but electrically neutral particles, such as neutrons, only the gravitational field enters as a factor in the determination of the trajectory; thus only gravitational interaction can be considered an element in the particle's space-time structure. In the case of particles with both charge and mass, however, both the electromagnetic and the gravitational field must be considered a factor in the determination of the particles' space-time structure.

Mass is obtained by the gravitational charge. We now express this concept of the determination of space-time structure with the following equations:

$$R_{ik} - \frac{1}{2} g_{ik} R = \frac{8G}{c^4} \, {}^g T_{ik} \tag{1}$$

Appendix 1

This is the equation of the gravitational field, while

$$R_{ik} - \frac{1}{2} g_{ik} R = \frac{8k}{c^4} \, {}^eT_{ik} \tag{2}$$

is the equation of the electromagnetic field, where ${}^gT_{ik}$ is the energy-impulse tensor of the gravitational field, ${}^eT_{ik}$ is the energy-impulse tensor of the electromagnetic field, G is a gravitational constant, and $k = 1/4\pi\varepsilon_0$ and ε_0 stand for the permittivity of the vacuum.

Combining the two equations we get

$$R_{ik} - \frac{1}{2} g_{ik} R = \frac{8 \, kG}{c^4(G+k)} \, (\pm {}^gT_{ik} \pm {}^eT_{ik}) \tag{3}$$

Linearity applies to weak fields only. This holds true of classical Newtonian fields as well.

Modification of the Tensor Equations of General Relativity

Einstein derived the celebrated tensor equations of general relativity by the variation of the effect-integral of gravitational field and gravitational mass. Thereby he obtained equations of motion, but regarded them formally as equations of space.

$$R_{ik} - \frac{1}{2} g_{ik} R = \frac{8 \, G}{c^4} \, T_{ik} \tag{4}$$

The questions is whether this coincidence is merely a matter of formalism. Could it be that the formal isomorphy between the equations of space and the classical equations of motion conceals a more fundamental relationship?

$$\delta(S_m + S_g) = \frac{c^3}{16G} \int (R_{ik} - \frac{1}{2} g_{ik} R - \frac{8 \, G}{c^4} T_{ik}) g_{ik} \, \overline{-g d} = 0 \tag{5}$$

Einstein varied the g_{ik} metric tensor which, according to David Hilbert, corresponds to the tensor of the gravitational potential. This potential has a velocity-dimensional aspect: m/s. The gravitational potential gives the flow velocity of the gravitational field at a given point. Field strength has an acceleration dimension: m/s², in other words, it is identical with the field's acceleration. Therefore gravitational force is identical with the mass-acceleration vector.

$$\vec{F} = m\vec{C} \qquad (6)$$

Regarding the internal structure of the Einstein equations we should note that

$$T_{ik} = \frac{2}{\sqrt{-g}} \left\{ \frac{\delta \Lambda \sqrt{-g}}{\delta g^{ik}} - \frac{\delta}{\delta x^l} \frac{\delta \Lambda \sqrt{-g}}{\delta \frac{\delta g^{ik}}{\delta x^l}} \right\} \qquad (7)$$

where $\Lambda = T - U$ is the Lagrange density of the matter-system, given that the L $(T - U)$ dV integral is a Lagrange-function.

$$R_{ik} - \frac{1}{2} g_{ik} R = \frac{1}{\sqrt{-g}} \left\{ \frac{\delta(C\sqrt{-g})}{\delta g^{ik}} - \frac{\delta}{\delta x^l} \frac{\delta(C\sqrt{-g})}{\delta \frac{\delta g^{ik}}{\delta x^l}} \right\}$$

C can be expressed with the Christoffel symbols

$$C = g^{ik} (\Gamma^m_{il}\Gamma^l_{km} - \Gamma^l_{ik}\Gamma^m_{lm}) \qquad (9)$$

Since the Christoffel symbols "appear" when the A^i vector is shifted in parallel within the coordinate system, C is an expression of space-time curvature.

$$\delta A^i = -\Gamma^i_{kl} A^k dx^l \qquad (10)$$

where δA^i is a transformation of the A^i vector.

It is significant that, according to Hilbert, the Christoffel symbol Γ^i_{kl} is a gravitational field strength tensor. Its dimension is m/s², i.e., it is acceleration. Λ, the Lagrange density, defines the density of the absorbing mass in space, that is, it expresses negative source density. Hence it marks the divergence of some vector space.

In turn, $\Lambda = div(\vec{g} \cdot a)$ where $\vec{g}$ is gravitational acceleration and a graviton density.

Consequently in the modified general relativity equations we have on the right-hand side the density of the absorber in space, and on the left-hand side an expression related to the space-time curvature generated by the absorber. It contains the expression for field strength (the Christoffel symbols). Field strength is none other, therefore, than the acceleration generated in the field by the effect of the absorber. As a result of this, acceleration macroeffects surface in the field, including the gravitational effect. These effects can thus be viewed as a consequence of the motion of massive particles in a field interpreted not as the formal construct of geometrical space-time, but as the physical vacuum. This interpretation renders the equations of general relativity consistent with the premises of Ψ field theory stated in chapter 5.

Postscript: Demonstration of the Unified Equations

$$\delta(S_m + S_g + S_q + S_e) = \int \left\{ \frac{c^3}{16G} \left[(R_{ik} - \frac{1}{2} g_{ik} R) \pm \frac{8G}{c^4} {}^g T_{ik} \right] + \frac{c^3}{16k} \left[(R_{ik} - \frac{1}{2} g_{ik} R) \pm \frac{8k}{c^4} {}^e T_{ik} \right] \right\} g_{ik} \sqrt{-g} d$$

Where S_m is the effect of gravitated mass,
S_g is the effect of gravitational field,
S_q is the effect of the charge,
S_e is the effect of the electromagnetic field,
G is the gravitational the constant,
$k = \frac{1}{4 \varepsilon_0}$, and ε_0 is the permittivity of the vacuum,
${}^g T_{ik}$ is the energy-impulse tensor of the gravitational field, and
${}^e T_{ik}$ is the energy-impulse tensor of the electromagnetic field.

From this we get:

$$\delta(S_m + S_g + S_q + S_e) = \int \frac{c^3}{16} \left[\left(\frac{R_{ik} - \frac{1}{2} g_{ik} R}{G} \pm \frac{8}{c^4} {}^g T_{ik} \right) + \left(\frac{R_{ik} - \frac{1}{2} g_{ik} R}{k} \pm \frac{8}{c^4} {}^e T_{ik} \right) \right] g_{ik} \sqrt{-g} d$$

And from this:

$$\frac{R_{ik}-\frac{1}{2}g_{ik}R}{G} + \frac{R_{ik}-\frac{1}{2}g_{ik}R}{k} = \frac{8}{c^4}(\pm {}^gT_{ik} \pm {}^eT_{ik}),$$

From this, multiplied by (G × k), we obtain:

$$R_{ik} - \frac{1}{2}g_{ik}R(G + k) = \frac{8Gk}{c^4}(\pm {}^gT_{ik} \pm {}^eT_{ik}),$$

From this, in turn, we have:

$$R_{ik} - \frac{1}{2}g_{ik}R = \frac{8Gk}{c^4(G+k)}(\pm {}^gT_{ik} \pm {}^eT_{ik}),$$

In this interpretation the structure of space-time, the physical vacuum, is determined by the gravitational and the electromagnetic fields.

Appendix 2

✦

Healing through the Ψ Field

Two Experiments

Maria Sági

In order to shed further light on the transpersonal coherence discussed in chapter 3, I report here on two experiments with remote healing. In these experiments I served as the healer and my method of diagnosis and therapy were tested.

The First Experiment

The first experiment was conducted by Günter Haffelder, director of the Institute for Communication and Brain Research of Stuttgart on June 3, 2001, on the occasion of a seminar of the

---

Maria Sági holds a Ph.D. in psychology from the Oetvös Lóránd University of Budapest and is an Associate Member ("Candidate") of the Hungarian Academy of Sciences. In addition to serving as a collaborator of the academy's Institute of Sociology, and program director of the Club of Budapest Foundation Hungary, she is the founder and director of the Koerbler Institute Hungary, with an active practice as a remote healer.

Hagia Chora Association in nearby Hohenwart (Germany). It was witnessed by about one hundred seminar participants and was protocolled both by Professor Haffelder and a volunteer, Dr. Heidrich Treugut. It was subsequently reported (in German) in the journal *Hagia Chora* (issue 9, August 2001).

At the beginning of the experiment I asked the subject—a fortyeight year-old volunteer from among the seminar participants—to give me a verbal report on his physical condition. He reported:

> Four years ago I turned to a doctor with a serious problem of articulation in my hands. I could hardly move my wrist, and the joints of my fingers were swollen and in pain. The doctor diagnosed multiple arthritis. I chose to try a natural therapy consisting of a vegetarian diet and a course of spiritual development. At present my finger joints and my right wrist move, but the left wrist is still stiff.

We then separated. The subject was seated in the main hall in the presence of Professor Haffelder and the seminar participants, and I moved to a distant room in the same building together with Dr. Treugut. Both the subject and I were wired with electrodes on our heads. The EEG instrument in the main hall, to which the electrodes were connected, displayed the electrical activity of our brains on two large monitors.

The steps in the experiment were as follows. First I tuned my brain and nervous system for receptivity to information under the given conditions, and then concentrated on the experimental subject. When I was satisfied regarding my own sensitivity, I begin to examine the subject. (I evolved a method originally developed by the noted Austrian researcher Erich Koerbler into an instrument for remote healing. It involves a one-arm dowsing rod and a system of geometric forms for focusing the healer's consciousness when receiving information from the patient, as well as when sending healing messages.) I first examined the subject's organs in sequence, and then his meridians. The colon showed symptoms that indicated a mild irregularity; I sought to correct this with the appropriate healing message. After this I found symptoms of inflammation in the left wrist, and for this,

Appendix 2

too, I sought the pertinent treatment. Among the meridians, that of the liver called for correction. Finally I examined the pancreas and corrected a mild malfunction there. When my reexamination indicated that no further meridian and organ correction was needed, I reinforced the balanced state of the subject with the use of a symbol derived from a method of healing traditionally practiced by Tibetan monks.

Since an important element in any such treatment is to determine the precise timing of the treatment, I undertook to do this during the last five minutes of the examination. The treatment for the colon and the left wrist was to last 10 days and that for the pancreas 6.5 days.

In reporting on the experiment, Haffelder noted,

> In general the process of examination between the healer and the subject occurs during a period of refined harmony between their respective brain activities, which manifests itself in the particular way that the frequency patterns of the two people influence each other. Thus high Delta activity (0–3 Herz) occurred in the healer, which shows the frequency of transmission of the nonverbal communication, as indicated by intensive alignment between healer and patient. On the part of the healer strong Delta activity manifested itself in the form of a significantly higher deviation of the regular rhythm in the range of 3–5 seconds in the left brain hemisphere. In alignment with the rhythm of this brain activity, Alpha and Delta activity occurred also in the patient. (...)
>
> The whole process of frequency activity during the treatment can be interpreted as a typical process that underlies our research on such phenomena and is in some cases reinforced. The healer inquires about the present condition of the patient, which is expressed through the activity of a certain frequency pattern. This image can be visualized by using a chronospectrogram. The healer balances the pattern and sends it back in a transformed form to the patient.

In my decade-long practice as a natural healer I have used this type of procedure in treating cases of acute inflammation, pains from injuries, and a variety of more serious maladies.

The Second Experiment

The essentially same experiment was repeated in Stuttgart at Professor Haffelder's Institute for Communication and Brain Research on October 29, 2001. It was monitored and subsequently documented by the spectrum-analytic method of the EEG recording the same as the Hohenwart experiment.

Before this experiment my forty-five year-old test subject described her complaint. She suffered from allergic bronchitis for approximately the past ten years, a condition that was especially excruciating in the morning hours. She consulted a number of physicians and received a great variety of medications, ranging from steroids and antibiotics to homeopathic remedies. I realized that there is not much point in concentrating on the symptoms themselves: I had to look for the causes. Thereafter the subject and I were wired with electrodes, and the subject went to another room in the laboratory, while I began the procedure for the diagnosis in the room shared by the experimenters. The electrical activity of our brains was displayed on monitors observed by the experimenters and was also recorded.

Following a preliminary examination of the patient, I proceeded to move back in the subject's life until I found a particular trauma that could account for her condition. This event occurred in the immediate postnatal period. I administered a healing message for the 10th minute after birth that was to last for a period of 2-minutes and 24-seconds, while the healing message for the seventeenth minute after birth was to last for 45 seconds. During the time that I administered these messages my brain exhibited EEG-waves in the slow Delta region. The brain of the subject replicated the patterns of my Delta waves with a delay of about 2 seconds. The effect was evident: as I was sending the healing messages the subject showed an aggravated form of her symptoms, coughing violently. When we concluded the experiment the coughing subsided and the subject calmed down.

On the 23rd of May of 2002 my test subject wrote:

> ... regarding my cough attacks, there were [after the experiment] some quieter periods and some periods in which I had violent coughing up to eight hours a day. Now it is quieter than ever before. The coughing did not entirely disappear, but it is within tolerable limits. In the last 10 years I never had such a quiet period as now.

Discussion

The Delta waves (0–3 Herz) that were shown by the EEG in these experiments are typical of normal deep sleep in adults, while Alpha waves (8–13 Herz) are typical of a restful state, usually with eyes closed. (In the normal waking state external stimuli generate Beta waves in the 20–30 Herz range.) It is significant, therefore, that in these experiments I had my eyes open and yet displayed electrical brain activity characteristic of deep sleep. Just as remarkably, the subjects displayed the same phenomenon. They sat relaxed, with closed eyes, but not asleep. In this state they proved receptive to the information I sent from my remote location, despite the absence of sensory connection between us.

The process of remote healing is generally the same. The information I receive as healer indicates the physical condition of the patient, including the nature of his or her complaint. The diagnosis may be as detailed as any personal examination in the doctor's office. I can find out if the patient suffers from a temporary malady or from a chronic illness. I can also identify the causes of the problem, and to what extent it is due to environmental conditions, such as electromagnetic or geomagnetic radiation, pollution, or other toxins.

I then proceed to determine the indicated therapy. If the therapy does not call for active cooperation on the part of the patient, I can "send" the healing information without the patient necessarily being conscious of it. Consciousness on the part of the patient is not a factor, however, provided she asks for, and accepts, the remote treatment. If she objects to it, her reception of the healing information may be blocked.

In the positive case the reception of the healing information by the patient affects her physical condition and can be verified through subsequent diagnosis, whether it is carried out through the remote method or by conventional means.

A significant feature of the information exchange between healer and patient is that it occurs independently of space. The healer can get the pertinent information even if the patient is in another town, in another country, or on another continent. The efficacy of remote healing is space-invariant. This was confirmed to me in a particularly striking manner in my work with the Psionic Medical Society of Great Britain. The members of the Society, reputed medical doctors, developed a remote-healing method using a "witness" from the patient (which can be any protein sample, usually a drop of blood or a strand of hair) for the diagnosis, and either direct healing messages or, more frequently, homeopathic remedies to effect the treatment (cf. chapter 3). Since 1994 I have been referring some of my cases in Hungary to one of the physicians in England; the patients would then receive the prescribed remedies by mail. I would follow the healing process through the periodic diagnosis of the patient's condition. An interesting phenomenon has come to light: the patients' recovery would usually begin at the time the treatment was determined, even though the remedy only reached them days later! The British physicians experienced the same phenomenon with most of their other patients.

Another significant feature of the diagnostic process is that information can be received not only regarding the momentary condition of the patient, but also regarding the patient's past history as it relates to her current condition. This is important, because in the case of chronic diseases it is often necessary to discover the original causes of the illness to effect healing. The experiment in Stuttgart testified to the feasibility of doing so.

I now give another example. A patient of mine, a man in his seventies suffered from neurodermitis for the past twenty years. He proved to be allergic to some seventy-two different foods, according to prior diagnosis by his doctors. I did not try to cure his allergies by working on the symptoms, but sought instead the originating causes. I found that he had suffered from some trauma when he was five weeks old, and that this event was connected with his present illness. I led him back to this stage of his life, and sent the rebalancing information using Tibetan symbols as well as Koerbler's geometric forms. During the treatment, which lasted about five minutes, the regressed patient cried aloud and waved his arms in the manner of an infant. Afterward he calmed down and became quiet. He accepted the dietary therapy I suggested and followed it rigorously. Three months later he wrote that the allergic spots and pains in his hands, arms, and feet had completely disappeared; it was his impression that he has fully recovered.

This case, together with scores of others in my decade-long healing practice, show that the information transferred during diagnosis and treatment is space- as well as time-invariant. Neither physical distance between healer and patient, nor elapsed time between the origins of the malady and its diagnosis, appear to limit it.

Conclusions

In a realistic context the phenomenon of remote healing suggests the presence of a holograph field as the physical information-transmitting medium. As Ervin Laszlo notes, in a holograph field every element of information is present at every point, and new information does not overwrite existing information but integrates with it by superposition. Consequently in a space- and time-invariant holograph medium information can be retrieved over any distance, from the present as well as from the past. This occurs regularly in the context of remote healing.

Phenomena of natural healing, and especially of remote healing, offer significant evidence that a nonlocal holograph field mediates the exchange of information between healer and patient. The Ψ field, it appears, is not an abstract theoretical construct but a working physical reality.

References

Aharonov, Y., and D. Bohm, 1959. "Significance of electromagnetic potentials in the quantum theory." *Phys. Rev,* 2nd series 115:3.

Akimov, A. E. and G. I. Shipov, 1997. "Torsion fields and their experimental manifestations." *Journal of New Energy* 2:2.

———, and V. Ya. Tarasenko, 1992. "Models of polarized states of the physical vacuum and torsion fields." *Soviet Physics Journal* 35:3.

Aspect, A., J. Dalibard and F. Roger, 1982. "Experimental test of Bell's inequalities using time-varying analyzers." *Physical Review Letters,* 49, 1804–1807.

Aspect, A. and P. Grangier, 1986. "Experiments on Einstein-Podolsky-Rosen-type correlations with pairs of visible photons," in: *Quantum Concepts in Space and Time,* R. Penrose and C. J. Isham, eds. Oxford: Clarendon Press.

Astin, J. A., E. Harkness, and E. Ernst, 2000. "The efficacy of "distant healing": a systematic review of randomized trials. *American Journal of Medicine* 132.

Backster, Cleve, 1968. "Evidence of a Primary Perception in Plant Life," *Int. Journal of Parapsychology* 10.4.

———, 1975. "Evidence for a Primary Perception at the Cellular Level in Plants and Animals," American Association for the Advancement of Science, Annual Meeting, 26–31 January 1975.

———, 1985. "Biocommunications Capability: Human Donors and *In Vitro* Leukocytes." *Int. Journal of Biosocial Research* 7.2.

Bajpai, R. P., 2002. "Biophoton and the quantum vision of life," *What Is Life?* Hans-Peter Dürr, Fritz-Albert Popp and Wolfram Schommers, eds. New Jersey, London, Singapore: World Scientific.

Barrow John D., and Frank J. Tipler, 1986. *The Anthropic Cosmological Principle.* London and New York: Oxford University Press.

Behe, Michael J., 1998. *Darwin's Black Box: The Biochemical Challenge to Evolution.* New York: Touchstone Books.

References

Beloussov, Lev, 2002. "The formative powers of developing organisms," *What Is Life?* Hans-Peter Dürr, Fritz-Albert Popp and Wolfram Schommers, eds. New Jersey, London, Singapore: World Scientific.

Benor, Daniel J., 1993. *Healing Research*, vol. 1, London: Helix Editions.

———, 1990. "Survey of spiritual healing research." *Contemporary Medical Research*, vol. 4, 9.

Bischof, Marco, 2002. "Introduction to integrative biophysics," *Lecture Notes in Biophysics*, in Fritz-Albert Popp and Lev V. Beloussov, eds. Dordrecht: Kluwer Academic Publishers.

Bohm, David, 1980. *Coherence and the Implicate Order.* London: Routledge & Kegan Paul.

——— and Basil Hiley, 1993. *The Undivided Universe.* London: Routledge.

Braud, W. G., 1992. *Human interconnectedness: research indications.* Revision 14:3.

———, and M. Schlitz, 1983. "Psychokinetic influence on electrodermal activity," *Journal of Parapsychology*, vol. 47.

Bucher, Martin A. and David N. Spergel, 1999. "Inflation in a Low-Density Universe," *Scientific American*, January.

Bucher, Martin A., Alfred S. Goldhaber, and Neil Turok, 1995. "Open Universe from Inflation," *Physical Review D*, 52:6 (15 September).

Buks, E., R. Schuster, M. Heiblum, D. Mahalu, and V. Umansky, 1998. "Dephasing in electron interference by a 'which-path' detector," *Nature*, vol. 391 (26 February).

Byrd, Randolph, 1988. "Positive therapeutic effects of intercessory prayer in a coronary care population," *Southern Medical Journal* 81:7.

Cardeña, Etzel, Steven Jay Lynn, and Stanley Krippner, 2000. *Varieties of Anomalous Experience: Examining the Scientific Evidence.* American Psychological Association, Washington, DC.

Chaboyer, Brian, Pierre Demarque, Peter J. Kernan, and Lawrence M. Krauss, 1998. "The Age of Globular Clusters in Light of Hipparcos: Resolving the Age Problem?" *Astrophysical Journal* 494:1 (10 February).

Chaisson, Eric, 2000. *Cosmic Evolution: The Rise of Complexity in Nature.* Harvard University Press, Cambridge.

Clayton, Philip D., 1997. *God and Contemporary Science.* Grand Rapids, Michigan: Eerdmans.

Conforti, Michael, 1999. *Field, Form and Fate: Patterns in Mind, Psyche and Nature.* Vermont: Spring Publications.

———, 1994. "Morphogenetic dynamics in the analytic relationship," *Psychological Perspectives* 30.

Del Giudice, E. G., S. Doglia, M. Milani, and G. Vitiello, in F. Guttmann and H. Keyzer (eds.), 1986. *Modern Bioelectrochemistry.* New York: Plenum.

References

Dobzhansky, Theodosius, *Genetics and the Origin of Species*, 1992. 2nd edition. New York: Columbia University Press.

Dossey, Larry, *Recovering the Soul: A Scientific and Spiritual Search*, 1989. New York: Bantam.

———, *Healing Words: The Power of Prayer and the Practice of Medicine*, 1993. San Francisco: Harper.

———, 1992. "Era III medicine: the next frontier." *ReVision* 14:3.

Duncan, A. J. and H. Kleinpoppen, 1988. "The experimental investigation of the Einstein-Podolsky-Rosen question and Bell's inequality," in Selleri, F. (ed.): *Quantum Mechanics versus Local Realism—The Einstein-Podolsky-Rosen Paradox*. New York: Plenum Press.

Dürr, S., T. Nonn and G. Rempe, 1998. "Origin of quantum-mechanical complementarity probed by a 'which-way' experiment in an atom interferometer" *Nature*, vol. 395 (3 September).

Einstein, Albert, Boris Podolski, and Nathan Rosen, 1935. "Can quantum mechanical description of physical reality be considered complete?" *Physical Review* 47.

Eldredge, Niles, 1985. *Time Frames: The Rethinking of Darwinian Evolution and the Theory of Punctuated Equilibria*, New York: Simon & Schuster.

———, and Stephen J. Gould, 1972. "Punctuated equilibria: an alternative to phylogenetic gradualism," *Models in Paleobiology*, edited by Schopf, Freeman, Cooper, San Francisco.

Elkin, A. P., 1942. *The Australian Aborigines*, Sydney: Angus & Robertson.

Frazer, Sir James G., 1890. *The Golden Bough: A Study in Magic and Religion*. 13 vols. London: MacMillan.

Frieden, Roy, 2001. "Physics from Fisher information," www.optics.arizona.edu/Frieden/fisher_information.htm.

Fröhlich, H., 1980. "Long range coherence and energy storage in biological systems." *Int. Journal of Quantum Chemistry* 2.

———, (ed.), 1988. *Biological Coherence and Response to External Stimuli*, Heidelberg: Springer Verlag.

Gazdag, László, 1998. *Beyond the Theory of Relativity*. Budapest: Robottechnika Kft.

———, 1989. "Superfluid mediums, vacuum spaces," *Speculations in Science and Technology*, vol. 12, 1.

Gilbert, S. F., J. M Opitz, and R. A. Raff, 1996. Resynthesizing evolutionary and developmental biology, *Developmental Biology*, 173, 357–372.

Goodwin, Brian, 1982. "Development and evolution," *Journal of Theoretical Biology*, 97.

———, 1989, "Organisms and minds as organic forms," *Leonardo*, 22, 1.

———, 1979. "On morphogenetic fields," *Theoria to Theory* 13.

Gould, Stephen J., 1983. "Irrelevance, submission and partnership: the changing role of paleontology in Darwin's three centennials, and a modest proposal for macroevolution," D. Bendall, (ed.), *Evolution from Molecules to Men*, Cambridge University Press.

———, 1991, Opus 200, *Natural History* (August).

——— and Niles Eldredge, 1977. "Punctuated equilibria: the tempo and mode of evolution reconsidered," *Paleobiology*, vol. 3.

Gribbin, John, 1993. *In the Beginning: The Birth of the Living Universe.* New York: Little, Brown & Co.

Grinberg-Zylberbaum, Jacobo M. Delaflor, M. E. Sanchez-Arellano, M. A. Guevara, and M. Perez, 1993. "Human communication and the electrophysiological activity of the brain." *Subtle Energies*, vol. 3, 3.

Grof, Stanislav, 1988. *The Adventure of Self-discovery*. Albany: State University of New York Press.

———, 2000. *Psychology of the Future.* "Albany: State University of New York Press.

———, 1996. "Healing and heuristic potential of non-ordinary states of consciousness: observations from modern consciousness research," mimeo.

———, with Hal Zina Bennett, 1993. *The Holotropic Mind.* San Francisco: Harper.

Guth, Alan H., 1997. *The Inflationary Universe: The Quest for a New Theory of Cosmic Origins.* California: Perseus.

Hagley, E. et al., 1997. "Generation of Einstein-Podolsky-Rosen pairs of atoms." *Physical Review Letters* 79 (1), 1–5.

Haisch, Bernhard, Alfonso Rueda, and H. E. Puthoff, 1994. "Inertia as a zero-point-field Lorentz force," *Physical Review A*, 49.2.

Hameroff, Stuart R., 1998. "Funda-Mentality": is the conscious mind subtly linked to a basic level of the universe?" *Trends in Cognitive Sciences*, 2.4.

Hansen, G. M., M. Schlitz and C. Tart, "Summary of remote viewing research," in Russell Targ and K. Harary, *The Mind Race.* 1972–1982. New York: Villard.

Haroche, Serge, 1998. "Entanglement, decoherence and the quantum/classical boundary" *Physics Today* (July).

Harris, W. S., M. Gowda, J. W. Kolby, C. P. Strycharz, J. L. Varck, P. G. Jones, et al., 1999. "A randomized control trial of the effects of remote, intercessory prayer on outcomes in patients admitted to the coronary care unit." *Arch. Intern. Med.*, 159.

Heisenberg, Werner, 1985. *Physics and Philosophy.* New York: Harper & Row.

———, 1975. "Development of concepts in the history of quantum theory." *American Journal of Physics*, 43:5.

References

Ho, Mae-Wan, 1993. *The Rainbow and the Worm: The Physics of Organisms.* Singapore and London: World Scientific.

———, 1996. "Bioenergetics, biocommunication and organic space-time," in *Living Computers,* ed. A. M. Fedorec and P. J. Marcer, The University of Greenwhich, Greenwhich, UK, March.

———, F. A. Popp and U. Warnke, eds., 1994. *Bioelectromagnetics and Biocommunication.* Singapore: World Scientific.

Hogan, Craig J., 1998. *The Little Book of the Big Bang.* New York: Springer Verlag.

Honorton, C. R. Berger, M. Varvoglis, M. Quant, P. Derr, E. Schechter, and D. Ferrari, 1990. "Psi -communication in the Ganzfeld: Experiments with an automated testing system and a comparison with a meta-analysis of earlier studies." *Journal of Parapsychology,* 54.

Hoyle, Fred, 1983. *The Intelligent Universe.* London: Michael Joseph.

———, G. Burbidge and J. V. Narlikar, "A quasi-steady state cosmology model with creation of matter," *The Astrophysical Journal* 410 (20 June 1993).

Kafatos, Menas, 1999. "Non-locality, foundational principles and consciousness." *Noetic Journal,* vol. 2 (January).

———, 1989. Bell's *Theorem, Quantum Theory and Conceptions of the Universe.* Dordrecht: Kluwer.

——— and R. Nadeau, 1990. *The Conscious Universe: Part and Whole in Modern Physical Theory.* New York: Springer Verlag.

Kaivarainen, Alex, 2002(a). "Unified model of Bivacuum, corpuscle-wave duality, electromagnetism, gravitation and time," *The Qui Manifesto,* Issue #1 (January), www.emergentmind.org/journal/htm.

———, 2002(b). "Unified Model of Bivacuum, Particles Duality, Electromagnetism, Gravitation and Time. The Superfluous Energy of Asymmetric Bivacuum," *The Journal of Non-Locality and Remote Mental Interactions,* 1, 3 (October).

Krauss, Lawrence M., 1998. "The End of the Age Problem and the Case for a Cosmological Constant Revisited," *Astrophysical Journal* 501:2 (10 July).

———, 1999. "Cosmological antigravity," *Scientific American* (January).

Laszlo, Ervin, 1987. *Evolution: the Grand Synthesis.* Boston: Shambala.

———, 1993. *The Creative Cosmos.* Edinburgh: Floris Books.

———, 1994. *The Interconnected Universe.* Singapore and London: World Scientific.

———, 1996. *The Whispering Pond.* Rockport, Shaftesbury, and Brisbane: Element Books.

Leslie, John, 1989. *Universes.* London and New York: Routledge.

———, (ed.) 1990. *Physical Cosmology and Philosophy*. New York: MacMillan.

Li, K. H., 1992. "Coherence in physics and biology," in F. A. Popp, K. H. Li and Q. Gu (eds.): *Recent Advances in Biophoton Research and its Applications*. Singapore: World Scientific Publishing.

———, 1994. "Uncertainty principle, coherence, and structures," in R. K. Mishra, D. Maass, and E. Zwierlein (eds.) *On Self-Organization*. Berlin: Springer Verlag.

———, 1995. "Coherence—a Bridge between micro- and macro-Systems," in *Biophotonics—Non-Equilibrium and Coherent Systems in Biology, Biophysics and Biotechnology*. L. V. Belousov and F. A. Popp (eds.) Moscow: Bioinform Services.

Licata, Ignazio, 1989. *Dinamica reticolare dello Spazio-Tempo* (Reticular dynamics of spacetime), Andromeda, Bologna, Inediti. No. 27.

Lieber, Michael M., 1998a. Environmentally responsive mutator systems: toward a unifying perspective. *Rivista di Biologia/Biology Forum*, 91.3.

———, 1998b. "Hypermutation as a means to globally repstabilize the genome following environmental stress," *Mutation Research, Fundamental and Molecular Mechanisms of Mutagenesis*, 421, 2.

———, Force and genomic change, *Frontier Perspectives*, 10,1 (2001).

Lorenz, Konrad, 1987. *The Waning of Humaneness*. Boston: Little, Brown & Co.

Mallove, Eugen F., 1988. *The Self-Reproducing Universe, Sky & Telescope* 76:3 (September).

Mandel, Leonard, 1991. *Physical Review Letters* 67:3, 318–321.

Mansfield, V., 1995. "On the physics and psychology of transference as an interactive field. "*http://lightlink.com/vic/field.htm.*

Maniotis, A., et al., 1997. "Demonstration of mechanical connections between integrins, cytoskeletal filaments, and nucleoplasm that stabilize nuclear structure," *Proceedings of the National Academy of Sciences*, USA, 4, 3.

Masulli, Ignazio, 1997. "Recurrences of form in the Old World as evidence of collective consciousness: a hypothesis for historical research," *World Futures* 48:1–4.

Maxwell, James C., 1873. *A Treatise on Electricity and Magnetism*. Oxford: Oxford University Press.

Michelson, A. A., 1881. "The relative motion of the earth and the luminiferous ether," *American Journal of Science*, vol. 22.

Montecucco, Nitamo, 2000. *Cyber: La Visione Olistica*. Rome, Italy: Mediterranee.

Morgan, Marlo, 1991. *Mutant Message Down Under*. Lees Summit, MM Co.

References

Nadeau, Robert, 1999. *The Non-Local Universe: The New Physics and Matters of the Mind.* Oxford: Oxford University Press.

Nelson, John E., 1994. *Healing the Split.* Albany: State Univeristy of New York Press.

Oschman, James L., 2001. *Energy Medicine: the Scientific Basis.* London: Harcourt.

Peebles, P. James E., 1993. *Principles of Physical Cosmology.* Princeton: Princeton University Press.

Penrose, Roger, 2000. *Shadows of the Mind: A Search for the Missing Science of Consciousness.* Oxford: Oxford University Press.

Perlmuter, S. G. M. Aldering, M. Della Valle et al., 1998. "Discovery of a Supernova Explosion at Half the Age of the Universe," *Nature,* vol. 391 (1 January).

Persinger, M. A. and S. Krippner, 1989. "Dream ESP experiments and geomagnetic activity," *The Journal of the American Society for Psychical Research,* vol. 83.

Prigogine, Ilya, J. Geheniau, E. Gunzig, and P. Nardone, 1988. "Thermodynamics of cosmological matter creation," *Proceedings of the National Academy of Sciences* USA, 85.

Primas, Hans, H. Atmanspacher, and A. Amman (eds.), 1999. *Quanta, Mind and Matter: Hans Primas in Context.* Dordrecht: Kluwer.

Psionic Medicine, Journal of The Psionic Medical Society and The Institute of Psionic Medicine. 2000. Vol. XVI.

Puthoff, Harold, and Russell Targ, 1976. "A perceptual channel for information transfer over kilometer distances: historical perspective and recent research," *Proceedings of the IEEE,* vol. 64.

Puthoff, Harold, 1987. "Ground state of hydrogen as a zero-point-fluctuation-determined state," *Phys. Rev. D,* 35:10.

———, 1989a, 1993. "Gravity as a zero-point-fluctuation force," *Phys. Rev. A,* 39:5.

———, 1989b. "Source of vacuum electromagnetic zero-point energy," *Phys. Rev. A,* 40:9.

———, 2001. "Quantum vacuum energy research and "metaphysical" processes in the physical world," *MISAHA Newsletter* # 32–35, January–December.

Radin, Dean, 1997. *The Conscious Universe: The Scientific Truth of Psychic Phenomena.* San Francisco: HarperEdge.

Rees, Martin, 1997. *Before the Beginning: Our Universe and Others.* New York: Addison-Wesley.

Rein, Glen, 1988. "Biological interactions with scalar energy-cellular mechanisms of action." *Proceedings of the 7th International Association of Psychotronics Research Conference,* Georgia (December).

———, 1989. "Effect of non-hertzian scalar waves on the immune system." *Journal of the U.S. Psychotronics Association*, 1, 15.

———, 1993. "Modulation of neurotransmitter function by quantum fields," in: Pribram, K. H., ed. *Rethinking Neural Networks: Quantum Fields and Biological Data*. Hillsdale NJ: Erlbaum.

———, 1998. "Biological effects of quantum fields and and their role in the natural healing process." *Frontier Perspectives* 7 (1).

Requard, Manfred, 1992. "From 'matter-energy' to 'irreducible information processing': arguments for a paradigm shift in fundamental physics," *Evolution of Information Processing Systems*, Kurt Hafner, ed., New York and Berlin: Springer Verlag.

Reyner, J. H., 2001. *Psionic Medicine: The Study and Treatment of the Causative Factors in Illness*. London.

Richards, T. L. and L. J. Standish, 2000. "EEG coherence and visual evoked potentials: Investigation of neural energy transfer between human subjects." Abstract No. 393. *Tucson 2000 Consciousness Conference*. http://www.imprint.co.uk/Tucson2000.

Riess, Adam G., Alexei V. Filippenko, Peter Challis, et al., 1998. "Observational Evidence from Supernovae for an Accelerating Universe and a Cosmological Constant." *Astronomical Journal*, 116:3 (September).

Rothe, Gunter M., 2002. "Electromagnetic, symbiotic and information interactions in the kingdom of organisms," *What Is Life?* Hans-Peter Dürr, Fritz-Albert Popp and Wolfram Schommers, eds. New Jersey, London, Singapore: World Scientific.

Rothman, Tony, and George Sudarshan, 1998. *Doubt and Certainty*. Reading, MA: Perseus Books.

Sági, Maria, 1998. Healing through the QVI field, *The Evolutionary Outrider: The Impact of the Human Agent on Evolution*. London: Adamantine Press and New York: Praeger.

Sakharov, A., 1968. "Vacuum quantum fluctuations in curved space and the theory of gravitation," *Soviet Physics—Doklady*, 12, 11.

Sarkadi, Dezső and László Bodonyi, 1999. "Gravity between commensurable masses." Private Research Centre of Fundamental Physics, *Magyar Energetika*, 7:2.

Schwartz-Salant, N., 1988. *The Borderline Personality—Vision and Healing*. Wilmette, IL: Chiron Publications.

Schwarzschild, B., 1998. "Very distant supernovae suggest that the cosmic expansion is speeding up." *Physics Today*, 51:6.

Selleri, F., ed., 1988. *Quantum Mechanics versus Local Realism—The Einstein-Podolsky-Rosen Paradox*. New York: Plenum Press.

Sheldrake, Rupert, 1981. *A New Science of Life*. London: Blond & Briggs.

———, 1988. *The Presence of the Past*. New York: Times Books.

Shipov, G. I., 1998. *A Theory of the Physical Vacuum: a New Paradigm.* International Institute for Theoretical and Applied Physics RANS, Moscow.

Smith, Cyril W., 1998. "Is a living system a macroscopic quantum system? *Frontier Perspectives* (Fall/Winter).

Steele, Edward J., R. A. Lindley, and R. V. Blandon, 1998. *Lamarck's Signature: New Retro-genes are Changing Darwin's Natural Selection Paradigm.* London: Allen & Unwin.

Steinhardt, Paul J. and Neil Turok, 2002. "A cyclic model of the universe," *Science,* vol. 296, 1436–1439.

Targ, Russell and Harold A. Puthoff, 1974. "Information transmission under conditions of sensory shielding," *Nature,* vol. 251.

——— and K. Harary, 1984. *The Mind Race.* New York: Villard Books.

Tart, Charles, 1975. *States of Consciousness.* New York: Dutton.

Taylor, R., 1998. "A gentle introduction to quantum biology," *Consciousness and Physical Reality,* vol. 1, 1.

Thaheld, F. H., 2001. "Proposed experiments to determine if there is a connection between biological nonlocality and consciousness," *Apeiron,* 8 (4), 53–66.

Tiller, William A., 1995. "Subtle energies in energy medicine." *Frontier Perspectives,* 4, 2 (Spring).

Tittel, W., J. Brendel, H. Zbinden, and N. Gisin, 1998. *Phys. Rev. Lett.* 81, 3563.

Tzoref, Judah, 1998. "Vacuum kinematics: a hypothesis," *Frontier Perspectives,* 7:2.

———, 2001. "New aspects of vacuum kinematics," *Frontier Perspectives,* 10.1.

Ullman, M. and S. Krippner, 1970. *Dream Studies and Telepathy: An Experimental Approach.* New York: Parapsychology Foundation.

Waddington, Conrad, 1966. "Fields and gradients," *Major Problems in Developmental Biology,* Michael Locke, ed. New York: Academic Press.

Wagner, E. O., 1999a. "Waves in dark matter," *Physics Essays* 12:1.

———, 1999b. "Structure in the Vacuum," *Frontier Perspectives* 10:2.

Welch, G. R., 1992. "An analogical 'field' construct in cellular biophysics: history and present status." *Progress in Biophysics and Molecular Biology* 57.

——— and H. A. Smith, 1990. "On the field structure of metabolic space-time," *Molecular and Biological Physics of Living Systems,* R. K. Mishra, ed. Dordrecht: Kluwer.

Wheeler, John A., 1984. "Bits, quanta, meaning," *Problems of Theoretical Physics,* A. Giovannini, F. Mancini, and M. Marinaro (eds.) Salerno: University of Salerno Press.

———, 1987. "Quantum cosmology," L. Z. Fang and R. Ruffini (eds.), *World Science*, Singapore.

Whitehead, Alfred North, 1929, 1978. *Process and Reality*. New York: MacMillan, revised edition New York: Free Press.

Whitteker, E. T., 1903. "On the partial differential equations of mathematical physics," *Mathematische Annalen* 57.

Wolkowski, Z. W., 1995. "Recent advances in the phoron concept: an attempt to decrease the incompleteness of scientific exploration of living systems." *Biophotonics*.

Wu, T. T. and C. N. Yang, 1975. "Some remarks about unquantized non-Abelian guage fields," *Physical Review D* 12:12.

Zeiger, Bernd F. and Marco Bischof, 1998. "The quantum vacuum and its significance in biology." Paper presented at The Third International Hombroich Symposium on Biophysics, Neuss, Germany, August 20–24 (mimeo).

Index

Abraham, Ralph, vii, viii
Acupuncture, 46
Adaptive mutation, 26
Aharonov-Bohm effect, 66
Akimov A. E., 62–63, 99
Altered states of consciousness, 33
Anthropic principle, 64
Aristotle, 51, 103
Aspect, Alain, 10

Backster, Cleve, 31
Backster's experiments, 31–32
Bajpai, R. P., 20
Bardeen, J., 58
Barrett, T. W., 66
Bearden, Thomas, 67
Behe, Michael J., 25
Beloussov, Lev, 21
Big Bang theory, 12, 16, 83, 84
Biophoton, 45
Bischof, Marco, 88–89, 100
Bivacuum, 63–64
Bodonyi, László, 61
Bohm, David, 74, 99
Bohr, Nils, 42, 111
"Boomerang" (Balloon Observations of Millimetric Extragalactic Radiation and Geophysics) project, 13, 83
Bose-Einstein condensate, 19
Brahman, 106
Brain-hemisphere synchronization, 32–33

Braud, William G., 37
Buks, Eyal, 8
Burbidge, G., 85
Byrd, Randolph, 35

C-value paradox, 21–22
Calabi-Yau space, 42
Casimir effect, 56
Casimir, H. G. B., 56
Cellular differentiation, 21
Chaisson, Eric, 96
Charged particles, 62, 67–70, 73–74, 80, 82, 88, 92–93, 99, 104
Christoffel symbol, 122
Closed universe, 12
COBE (Cosmic Background Explorer), 12, 13
Coherence
  anomalous, 1, 3ff, 96
  explained, 79–93
  in the living world, 17
  in the mind, 27–38
  in the physical world, 5–16
  intra-organic, 18, 22
  reduction, 114
  transorganic, 22–26, 88, 107
  transpersonal, 29–38, 125
  understanding of, 39–48
Coherence-space, 11
Coherence-time, 11
Complementarity principle, 111–112
Concepts of Divine, 114–118
Conforti, Michael, 33

Connectivity hypothesis, 49–77, 97, 100
  postulates of, 65–71
Connectivity, philosophical implications of, 110–118
Consciousness research, 27–28
Cooper, L. N., 58
Copenhagen interpretation, 6
Cornell, Eric A., 19
Correlation, 40–41, 79
  degree of, 75–77
Cosmic macrostructures, 82
Cosmic plenum, 51–57, 65–71, 79, 100, 106, 112, 115, 118
  energy density of, 53
Coyle, Michael J., 64

DASI (Degree Angular Scale Interferometer), 13, 83
Davies, Paul, 57
Deism, 116
Delta activity (EEG), 127
Democritus, 51
Dirac, Paul, 13, 54, 56
Dirac-sea, 55–56
Distant healing (*also* remote healing), 37–38, 129–131
Divine intervention, 116
Dobzhansky, Theodosius, 24
Dossey, Larry, 37
Double-slit experiment, 6
Downward causation, 107
Dynamic order, 17–18

Eddington, Sir Arthur, 13
EEG (electroencelograph) waves, 126–128
Einstein, Albert, 9, 91, 99, 119, 120, 121, 122
Eldredge, Niles, 24
Embryogenesis, 44
Entropy, 18
Enz, Charles, 58
EPR experiment, 9–10
Ether, 52, 99
Evolution
  cosmic, 107–109
  interactive, 106–107, 111
Evolutionary biology, 22–23

Faraday, Michael, 41
Feynman, Richard, 11
Fields, 40–48, 49, 97
  biological, 43–47
  classical, 41
  morphogenetic, 44
  physical, 42–43
  transpersonal, 47–48
Figure-ground switch, 46–47
Fisher information, 98
Fourier, Jean Baptiste, 68
Frazer, Sir James, 36
Free-energy density, 96
Freedom, 112–113
Fresnel, Jacques, 52
Frieden, Roy, 98
Fröhlich, Hans, 19

Gazdag, László, 57, 60, 99, 119–124
Geheniau, J., 85
Gene-number paradox, 21–22
General Relativity, equations of, 119–124
Genetic determinism, 21
Gestalt Psychology, 28–43
Gibbs, Josiah, 56
Gilbert, Scott F., 45
God, 115, 116, 118
  consequent nature of, 116–117
  primordial nature of, 116
Goodwin, Brian, 45
Gould, Stephen Jay, 24
Gravitational field, 92, 121, 124
Gravity, 60, 61, 85, 92, 119
Gravity constant, 14
Grinberg-Zylberbaum, Jacobo, 30
Grof, Stanislav, 33
Gunzig, A., 85
Gurwitch, Alexander, 44
Guth, Alan H., 15

Haffelder, Günter, 125, 126, 127, 128
Haisch, Bernhard, 59, 60

Index

Hameroff, Stuart, 48
Healing (*see also* distant healing), 125–131
  information, 129–131
  first experiment, 125–128
  second experiment, 128–129
Heaviside, Oliver, 66
Heisenberg, Werner, 6
Hertz, Heinrich, 66
Heyblum, Mordechai, 8
Hilbert, David, 121, 122
Hindu cosmology, 106
Ho, Mae-Wan, 18
Holography, 70
Horizon problem, 15
Hoyle, Sir Fred, 25, 85

Idealism, 110–111
Ignatovich, V. K., 67
Inertia, 59, 60
Inflation theory, 15–16, 83
Information (*also* in-formation), 74–77, 80–93, 97–100, 104, 131
  transcyclic, 87
Instinct, 90
Integral science (*also* integral quantum science), 2, 27, 73, 95–101
Introspective observer, 117
Irreducibly complex system, 25
Isolation of genome, 26

Jánossy, Lajos, 57

Kafatos, Menas, 14, 82
Kaivarainen, Alex, 63, 64
Ketterle, Wolfgang, 19
Koehler, Wolfgang, 43
Koerbler, Erich, 38, 126

Lamb-shift, 56–57
Lamoreaux, S. K., 56
Langs, Robert, 33
Laws of physics, 98
Li, Ke-Hsuih, 11
Licata, Ignazio, 59

Lieber, Michael M., 25
Linde, Andrei, 15
Living organism, 18–22
Living state, 19
Local universes, 108–109

Macroscopic quantum system, 19–20
Macroworld hypothesis, 82–87
Mandel, Leonard, 8
Manifest domain, 103–109
Manifest entities, 104–109, 117
Maniotis, A., 95
Masulli, Ignazio, 34
Materialism, 110–111
Matter, 104, 111
Matter-density, 12
MAXIMA (Millimeter Anisotropy Experiment Imaging Array), 13, 85
Maxwell, Clerk, 41, 51, 66
Mental potential, 112
Mesoworld hypothesis, 87–91
Metaphysics of connectivity, 103–109, 111
Metaverse, 16, 85, 108, 109, 115, 118
Michelson, A. A., 52
Michelson-Morley experiment, 52
Microworld hypothesis, 80–82
Montecucco, Nitamo, 32, 33
Morality, 113–114
Murphy, Nancey, 115

Nadeau, Robert, 82
Nardone, P., 85
Narliker, J. V., 85
New homeopathy, 38
Nonlocal medicine, 37
Nonlocality
  cosmic, 11–16
  quantum, 5–11
Nonsensory information, 47, 112

Occam's Razor, 73, 116
Open universe, 12
Opitz, John M., 45
Oschmann, James, 46

Paradigm-shift, 1, 40, 95
Particle-mass, 55
Peacock, Arthur, 115
Penrose, Roger, 16, 48, 63, 99
Physical space-time, 59
Physical vacuum, 62, 119–124
Physis, 111
Phyton, 62–63
Planck constant, 14, 61, 62
Planck, Max, 54
Planck-time, 15
Plato, 105
Polkinghorn, John, 115
Popp, Fritz-Albert, 19, 45, 99
Prayer, intercessory, 35
Pre-space fluctuation, 84
Prigogine, Ilya, 85
Primas, Hans, 99
Psi (Ψ) field, 68–77, 80–91, 119, 131
Psionic Medical Society, 130
Psionic medicine, 36
Psyche, 111
Punctuated equilibrium, 24
Puthoff, Harold F., 29, 59, 60, 99, 100

Q-tick, 6
QSSC (Quasi-Steady State Cosmology), 85
Quantum phenomena, 5–11
Quantum potential, 99
Quantum vacuum (*see also* cosmic plenum), 12, 51

Radin, Dean, 36, 37
Raff, Rudolf A., 45
Random mutation, 23–25
Rein, Glenn, 99
Relativistic effects, 58–59, 123
Remote viewing experiments, 29–30
Requard, Manfred, 59
Rothe, Günter, 45
Rueda, Alfonso, 59

Sági, Maria, 38, 125–131
Sakharov, Andrei, 57

Sarkadi, Dezső, 61
Scalar field, 68, 70, 93
  capacity of, 72
Scalar potentials, 65–67
Scalar waves, 69
Schlitz, Marilyn, 37
Schrieffer, G. R., 58
Schrödinger, Erwin, 10, 17
Scientific revolution, 95–96
SED (stochastic electrodynamics), 54
Sheldrake, Rupert, 45
Shipov, G. I., 62, 63
Space, 51–55
Species-formation, 24
Split-beam experiment, 6–7
Standard Model (of particle physics), 55–56
Steinhard, Paul J., 83, 85
Steinhardt-Turok model, 83, 85–86
String theory, 42–43
Superfluid helium, 58
Superfluid vacuum, 57, 58
Sympathetic magic, 35–36

Targ, Russell, 29
Telesomatic effects, 35–38
Tesla, Nicola, 66, 67
Theological perspectives, 114–118
Tiller, William A., 67
Transdisciplinary unified theory, 2, 17
Transferred potentials, 30
Transpersonal
  contact, 34
  experiences, 47
  transference, 30
Treugut, Heinrich, 126
Tsoref, Judah, 62
Turok, Neil, 83, 85
Two domains of reality, 103–106
Two-fluid hydrodynamics, 58

Unified equations, 123–124
Unified theory, 97
Unified vacuum, 55

Universal constants, 14
Unruh, William, 57
Upward causation, 107

Vacuum fluctuations, 16
Vacuum Kinematics, 62
Vacuum-particle interaction theories, 57–64
Virtual domain, 104–109, 114

W-waves, 64
Waddington, Conrad, 44
Wagner, O. E., 64
Wavefunction (Ψ function), 68–71, 80–91
Wavefunction, macroscopic, 88–89
Weinberg, Stephen, 56
Weiss, Paul, 43, 44

Wheeler, John A., 6, 7
Which-path detection, 8–9
Whitehead, Alfred North, 104, 105
Whittaker, E. T., 69
Wieman, Carl E., 19
Wigner, Eugen, 6
WMAP (Wilkinson Microwave Anisotropy Probe), 13
Wu, T. T., 66

Yang, C. N., 66
Young, Thomas, 6

Zeiger, Bernd, 88, 89
Zhang, Chang-Lin, 46
Ziolkowski, R. W., 67
ZPE (zero-point energy), 54
ZPF (zero-point field), 55, 57, 59, 60